P9-DYO-305

CECIL COUNTY
PUBLIC LIBRARY

DEC 2 7 2016

301 Newark Ave
Elkton, MD 21921

IF OUR BODIES COULD TALK

IF OUR BODIES COULD TALK

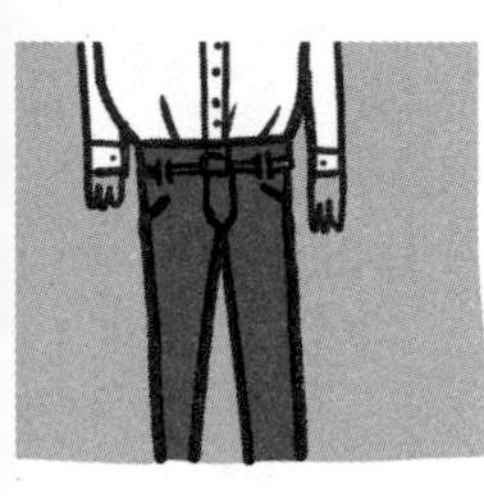

A GUIDE TO OPERATING AND MAINTAINING A HUMAN BODY

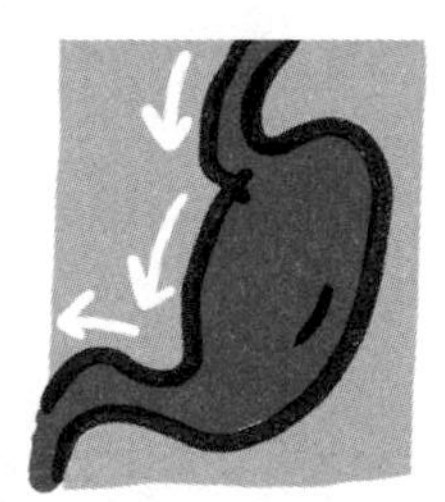
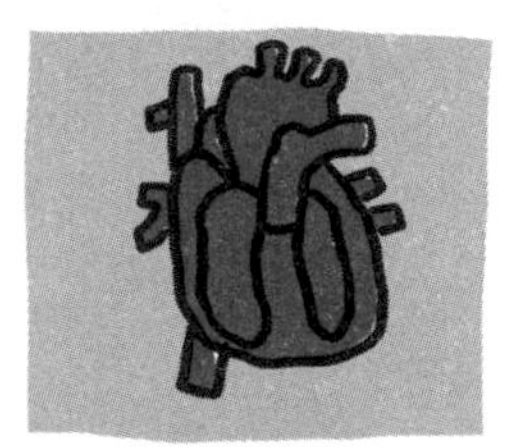

JAMES HAMBLIN

DOUBLEDAY

NEW YORK LONDON TORONTO SYDNEY AUCKLAND

The information and advice presented in this book are not meant to supersede or substitute for the expertise of a personal physician. Consult a health care professional before acting on any of the information presented or implied in these pages.

Copyright © 2016 by James Hamblin

Illustrations copyright © 2016 by Hallie Bateman

All rights reserved. Published in the United States by Doubleday, a division of Penguin Random House LLC, New York, and distributed in Canada by Random House of Canada, a division of Penguin Random House Canada Limited, Toronto.

www.doubleday.com

DOUBLEDAY and the portrayal of an anchor with a dolphin are registered trademarks of Penguin Random House LLC.

Book design by Iris Weinstein
Jacket image: heart shape, gst/Shutterstock
Jacket design by John Fontana

Library of Congress Cataloging-in-Publication Data
Names: Hamblin, James, author.
Title: If our bodies could talk : a guide to operating and maintaining a human body / James Hamblin, M.D.
Description: First edition. | New York : Doubleday, [2016]
Identifiers: LCCN 2016046694 | ISBN 9780385540971 (hardback) | ISBN 9780385540988 (ebook)
Subjects: LCSH: Health—Popular works. | Health—Miscellanea. | BISAC: HEALTH & FITNESS / Healthy Living. | SCIENCE / Life Sciences / Human Anatomy & Physiology. | HEALTH & FITNESS / Reference.
Classification: LCC RA776.H183 2016 | DDC 613—dc23
LC record available at https://lccn.loc.gov/2016046694

Manufactured in the United States of America

10 9 8 7 6 5 4 3 2 1

First Edition

To Sarah Yager, John Gould,
and the rest of *The Atlantic*

CONTENTS

TWO • PERCEIVING: The Feeling Parts

THREE • EATING: The Sustaining Parts

FOUR • DRINKING: The Hydrating Parts

FIVE • RELATING: The Sex Parts

PROLOGUE

My medical school roommate became an ophthalmologist and moved to Texas. He encouraged me to address here the most common question that people ask him in conversation when they learn his occupation:

If I lose a contact lens in my eye, can it get into my brain?

I laughed. He didn't. The question has lost its humor to him.

There are common, debilitating eye diseases that people could be asking him about—like macular degeneration, or night blindness, or glaucoma, which will affect 112 million people by 2040, leaving many without sight.

That last one resonates with me because I have it. The pressure inside my eyeball is higher than it should be. My eye won't explode—though that image does absurdly haunt me. The decline in most glaucoma is, rather, insidious. I'm told I won't even notice as my eyes "fail"—a term that doctors use commonly, unthinkingly, until we're the ones whose parts are failing. It's more accurately the case that the pressure inside my eyes will gradually damage the dense focus of nerves in the retina at the back of my eye. I stand to slowly lose vision at the peripheries of my field of view, and then entirely.

This won't happen for *years*.

Which is all to say that we have our reasons for concerns about our eyes and other parts. Every one is valid. Sometimes it helps to put them in the context of other people's problems—of

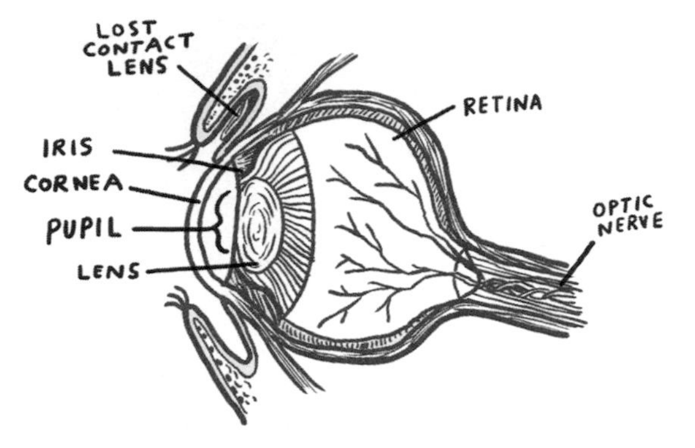

how bad things could be. But sometimes it doesn't. So, to the question at hand: The space under our eyelids does not connect to our brains. It's a cul-de-sac that ends about halfway over our eyes. Our brains are safe from contact lenses.

This anatomy is something you might have seen if you were among the forty million people who experienced the most popular traveling museum exhibition of all time: Body Worlds. Though you may have missed the cross sections of human heads if you were too distracted by the corpses that were posed having sex. That element of the exhibition shocked many attendees, as did rumors about dubious procurement of the bodies. Maybe most shocked was the art world, though, by the very fact of the enormous and enduring popularity of an exhibit that might have been alternatively titled Actual Corpses.

Of all the art bestowed to us throughout history, why should what is essentially a glorified biology lab be so successful and adored? Especially as most of us are otherwise so averse to discussing so much of what our bodies do, and considering death, in any realistic way?

Body Worlds is the brainchild of the German anatomist Gunther von Hagens, who invented the "plastination" process that allows bodies to be preserved without decomposing. While most exhibits come and go, Body Worlds has now been appearing around the world for more than two uninterrupted decades. It even stays open on Friday nights to accommodate couples who want to make a date of it.

The marketing professor Kent Drummond at the University of

Wyoming surmises that Body Worlds speaks to people because it manages to juxtapose our distaste for the abject with our desire to live forever. The displays draw on the sublimity of our mortality without leaving us feeling consumed by it. Drummond came to this understanding by studying not only the corpses but also living people as they move through the exhibition. He writes in a field note, "In one oft-repeated pattern of interaction, a man points to a body part in a casement, then explains to a woman how that part functions. He does so by showing her where it is on his body."

This masculine exhibitionism may be more sobering than the corpses themselves. It's also in keeping with the grand vision of Von Hagens, who identifies as a "medical socialist." He believes that health information should be a social good: the constellation of factors that go into picking up a cigarette should include an abiding familiarity with a black, necrotic, emphysematous lung, which should not be exiled to textbooks and morgues. Inside Body Worlds, we can clearly see and contemplate our organs and their mortality, if only for one date night. Placards scattered throughout the exhibition urge self-reflection, such as Kahlil Gibran's "Your body is the harp of your soul."

I don't think that means anything, but Von Hagens's philosophy does. Democratizing health information has become the norm well outside the walls of his exhibit. A past world in which doctors were the keepers of all medical knowledge—whose job was primarily dispensing directives—is gone. Most of us are rather awash in information now—so much so that it can be difficult to know what to make of it.

Googling bodily concerns isn't always helpful. Anonymous people in forums can be found debating all things—including the great dilemma of the contact lens. (Can it get far into my eye to my brain and cause damage? What if it goes down my spine and into my shoe? Is it still safe to wear again?) And even when you find a reliable-looking source of health information, there is almost always a ready conspiracy theorist who writes with passion and anecdotal logic to warn everyone not to trust that source. Usually it's a guy on Reddit named Gene. Gene personally lost five hundred contacts in his brain, and he had to have them surgically removed. He has the mass of desiccated contact lenses sitting in a jar on his desk.

While contact lenses can't get into our brains, they can in rare cases get stuck in the cul-de-sac above or below the eyeball. Like most anything that gets stuck in our bodies, this can be a source of infection. The pus around the contact lens can drain into a person's sinuses and spread into their pharynx. This has happened to me. I thought my contact had just fallen out, but no. Six days later, it came out. In the interim, I got pretty sick.

So, do seek medical attention if you have a trapped contact that is refusing to come out. (I hope everyone read this whole answer and not just the beginning.)

At Stanford University, the gaunt, bespectacled professor Robert Proctor teaches a course called "History of Ignorance." If he believed ignorance were simply the absence of knowledge, cured by imparting facts, his course would be dull. Instead, he argues that ignorance is the product of *active cultivation*. It spreads through marketing and through rumor, and it spreads much more easily than wisdom.

Contrasting his idea with the study of knowledge, epistemology, Proctor has named the study of ignorance *agnotology*. The word is still not in the *Oxford English Dictionary,* though it is relevant to the dictionary's most recent Word of the Year, 😂.

In 1977, a wide-eyed Proctor left home in Indiana to pursue a graduate degree in the history of science at Harvard. There he found himself "disturbed and puzzled" at the apathy that his professors had for "what ordinary people think." He saw it partly as elitism; the rest was a darker sense of futility. "At the time, half of the country thought Earth was six thousand years old," he recalls. That's off by about 4.6 billion years. But more puzzling to Proctor than the discrepancy itself was the apathy toward it on the part of his colleagues in academia. So he decided that someone should study "what people don't know and why."

The classic example of purposeful ignorance is that created by the tobacco industry. Ever since tobacco was clearly proven to cause lung cancer in the 1960s, the industry has attempted to cultivate doubt in science itself. It cannot refute the facts of cigarettes, so it turned the public opinion against knowledge. Can anything *really* be known?

The strategy was brilliant. Proctor calls it "alternative causation," or simply, "experts disagree." Tobacco companies didn't have to disprove the fact that smoking causes cancer; all they had to do was imply that there are "experts" on "both sides" of a "debate" on the subject. And then righteously say that everyone is entitled to their belief. The tactic was so effective that it bought the industry decades to profit while reasonable people were uncertain if cigarettes caused cancer.

As Proctor put it, "The industry knew that a third of all cancers were caused by cigarettes, so they made these campaigns that would say experts are always blaming *something*—brussels sprouts, sex, pollution. Next week it'll be something else."

Once you start looking for this tactic, it's hard to miss. It's nowhere more common than in the messages about our bodies. Proctor rattles off examples: vaccine agnotology, clitoral agnotology, food agnotology, milk agnotology. He likes to say we live in "the Golden Age of Ignorance." Because of the way information flows, "powerful agencies are able to create ignorance and spread lies through more vehicles than there have ever been in history."

Proctor is not alone in this thought: There's undeniably more scientific misinformation and marketing-based "facts" about our bodies coming at each of us daily than in entire lifetimes of generations past. And as we increasingly read only the articles that appear in our inboxes and curated social media feeds, "it's easier and easier," Proctor says, "to silo yourself into a tunnel of ignorance."

To allow ourselves to be challenged—to welcome it, and to seek it out—is to guard against the purposeful cultivation of ignorance. To be a doctor today is ever closer to its Latin root, *docere* ("to teach"), which I take to mean sharing habits of thought. The challenges for doctors and patients alike are in contextualizing, separating marketing from science, finding the lines between known and unknown, and discerning the motives of the people who attempt to define and redefine health and normalcy. If we all equip ourselves accordingly, we might reckon with the onslaught of bodily messages and maintain a solid understanding of ourselves that allows us to relate productively to one another and move cogently, even happily, through the world.

So this book is a practical approach to understanding our bodies, predicated on the idea that memorizing facts is less important than developing insight. This is also a corrective to the approach that drove me away from practicing medicine. In premed courses, throughout medical school, and in the three years that I spent in residency, I memorized roughly infinity facts. During that time, it wasn't uncommon for people teaching me to admit that I just needed to memorize these things to pass a test, that no doctors in the real world actually remember all this stuff—all the structures of amino acids, and the names of the small arteries that supply the elbow, and every possible minor side effect of every known medication. These are things that can be easily looked up at any moment. But still, on the exams that propel a person to success in the field, minutiae are currency.

After years of memorization, the overall effect was one of jumping through hoops with the explicit purpose of getting to the next hoop. My mentors advised me that if I didn't love the process by that point, I probably wasn't going to love the end result. So in 2012 I went on leave from the radiology residency at UCLA. I got the opportunity to take a job as editor of the health section of *The Atlantic*'s digital magazine—a publication I had always read and loved. I was happier and more engaged, learning in a way that made sense to me.

So I resigned from UCLA. I justified leaving a very stable, lucrative career for a very unstable industry by the fact that there are not enough science journalists or doctors working in public health. I wanted to have an impact on the roots of problems more than the symptoms, to question the textbooks rather than memorize them, and, ideally, to make people laugh. Journalism allows me to have some hand in public scientific literacy, and that might be, I mean to suggest with this book, the most valuable tool in pursuing health and happiness.

I've yet to regret my decision for any extended period of time.

The book began as a collection of straightforward answers to common questions about bodies, because, professionally and personally, I get a lot of those. It grew to interrogate those questions—why we care, or don't, about how our bodies work, and how our understandings of our bodies shape how we come to believe what's to be done about ourselves. At the root of the most virulent diseases and violent mistreatment of one another is ignorance, and much of that begins with fundamental misunderstandings of our differences—understandings of ourselves and others that begin with our bodies. The questions in this book often began with little more than minor bodily curiosities, which, looked at more closely, are not at all minor.

Many of the answers are, rather, stories about why we don't have concrete answers. Sometimes the most interesting thing is knowing why we don't know, and the point is in the considering, and being comfortable in not knowing. Health is a balance between acceptance and control.

What is normal?

Too many daily decisions about what to put into our bodies—how to put what in where, and what to do with that body once it's full of things—come down to vague ideas of what is good or bad, healthy or unhealthy, natural or unnatural, self or other. In a world of inordinate complexity, we instinctively attempt to put things into these binary categories.

University of Pennsylvania psychologist Paul Rozin believes that we do this to help maintain a sense of order. He calls this instinct "the monotonic mind." Even though we know better, we tend to resist the idea that most things are beneficial in some contexts or amounts, and harmful in others. It's easier to regard things as simply bad or good, to be adored or avoided.

In that tendency to seek order and control, an abiding theme among bodily questions and concerns is the concept of *normal*. The word tends to mean different things to scientists, who use it in every other sentence, with statistical deviation in mind, and nonscientists, who are more likely to hear in it judgment.

Is it normal that I can bend my finger all the way back until it touches my wrist? Statistically, no. That doesn't mean it has implications for your health.

Maybe more consequential than its normalcy is the simple fact that if you're able to do that, you probably also know that people don't like watching it. The Canadian psychologist Mark Schaller argues that we're wired to be averse to looking at things like people flipping their eyelids inside out—to say nothing of injuries like bones broken or blood outside of vessels—because of a concept called the "behavioral immune system." We're repelled because on some level we sense a threat to our health.

Clearly, if our reactions to eyelid flipping or finger flexibility are an indication, our behavioral immune system is far from perfect at acting on only credible threats. Schaller has implicated this faulty self-preservation instinct in all sorts of behaviors, which lead us to isolate ourselves into cliques and communities based on the appearances and functions of our bodies.

At a grander scale, then, the system he proposes can be seen to be involved in many of the world's fundamental divides (racism, ageism, xenophobia). It stems from our understanding of ourselves—which, again, begins with our bodies. Understanding oneself as *abnormal* can register anywhere between liberating and suffocating.

Or the idea of "normal" can be rejected altogether. A central tenet of the Deaf community, for one, is that deafness should not be considered a disease to be treated or cured. The community does not consider people to be hearing "impaired" and rejects any reference to hearing "loss." The same is true for some other communities long marginalized by some bodily process.

Still, even as normalcy is a loaded concept, it's sometimes a necessary lens through which to understand disease and, ultimately, to reduce suffering. Identifying outliers is central to the study and improvement of health. Science can't skirt the concept of normalcy, and neither can this book. But I do my best here to separate statistically common ways of being from judgments of value, of right or wrong, of implying that there is some ideal way to operate, look, feel, or be.

What is health?

At the founding of the World Health Organization in 1948, the group's constitution defined *health* in a way at once obvious and radical: "A state of complete physical, mental, and social well-being, and not merely the absence of disease or infirmity."

With that, the World Health Organization hoped to inspire a new purview for the medical profession.

It failed. In much of the world, "health care" systems still today focus exclusively on the absence of disease or infirmity.

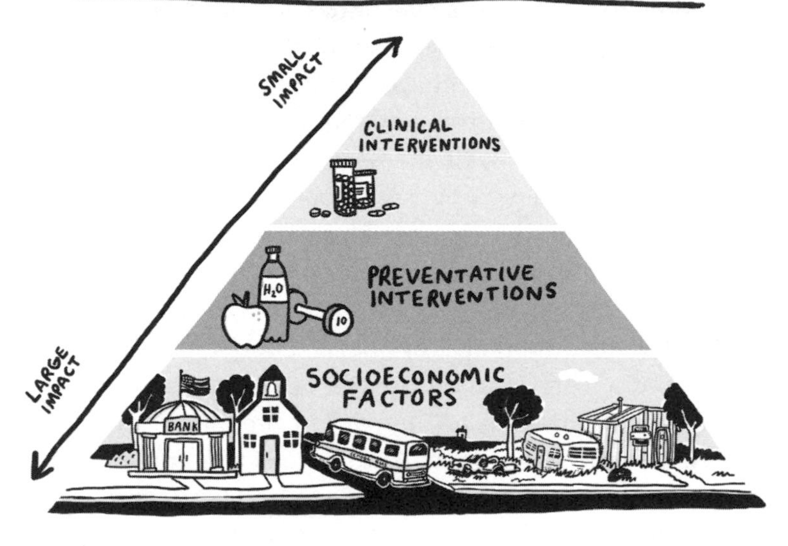

More specifically, they focus on treating disease after it already exists. In the last few years, though, an upheaval has begun.

The spring of 2015 saw the swearing-in ceremony of Vivek Murthy, who was quickly among the most controversial U.S. surgeons general. Conservative politicians attempted to block his appointment because of a tweet three years prior. Murthy had written, "Tired of politicians playing politics [with] guns, putting lives at risk [because] they're scared of NRA. Guns are a health care issue."

It wasn't even an especially revelatory tweet. Homicide and suicide are perennially among the leading causes of death in the country, a fact that has recently led the American Medical Association and other physician bodies to recommend that doctors ask all patients as a standard screening question—just as they should ask if patients wear seat belts and have fire extinguishers—if they keep guns in their homes. But it was the sort of tweet that can keep a person out of political office, in a country where the National Rifle Association and its elected officials have forbidden the Centers for Disease Control and Prevention from even studying gun violence.

After a harrowing welcome to politics, Murthy ultimately made it through confirmation. When he took the podium to be sworn in, he spent little time talking about the traditionally paradigmatic doctorly pursuits—treating pancreatitis, performing colectomies or cardiac ablations. Actually, no time on those things. He underscored instead how preventable illness influences and is influenced by education, employment, the environment, and the economy. He called for the building of "the great American community" that will approach health as a unified endeavor.

His words build on a growing movement in the medical profession. While the United States spends the most money per person on health care of any country, it ranks forty-third in life expectancy. And more important than longevity, the United States is near the bottom of the ranking list among wealthy countries in personal health status. In a pivotal 2007 paper in the *New England Journal of Medicine,* physician Steven Schroeder argued that medical care accounts for only about 10 percent of what determines a person's likelihood of dying young. Genetic factors might account for another 30 percent or so. The remaining 60 percent came down to social and environmental circumstances and behaviors. These are necessarily rough estimates, but they serve to push back against the way of thinking that leads us to think about hospitals, pills, and procedures when we think about improving health. Schroeder argues in the journal, "Even if the entire U.S. population had access to excellent medical care—which it does not—only a small fraction of [premature] deaths could be prevented."

This is not to say that modern health care cannot accomplish amazing things in treating diseases, some of which I'll get into in these pages, but that we rely too heavily on a mind-set where

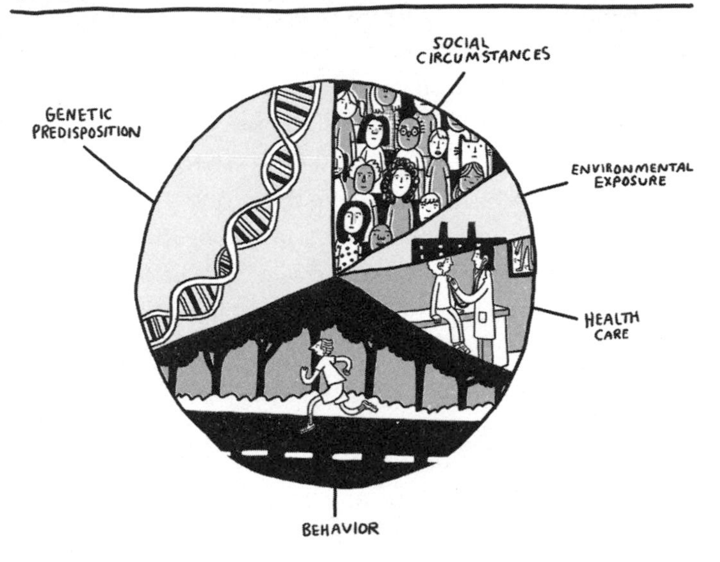

our system fixes problems, and not heavily enough on creating systems where these problems do not arise.

Over the decades, physicians have tended ever more toward specialties (and subspecialties, and sub-subspecialties) that treat discrete organ systems—dermatologic oncology, pediatric autoimmune gastroenterology, neuro-oncology, and so on—which has been critical to managing the wealth of information as science advances. But it has also left behind comprehensive approaches to the conditions that sicken and kill most people, first among them being the disease that we vaguely call "metabolic syndrome." This manifests as a combination of obesity, diabetes, and cardiac death. This is primarily a disease of society, a disease of life.

As patients, the concept can be liberating: Our control over our health is great. And, more interesting, our ability to improve the health of others is great.

A typical textbook of anatomy and physiology is still today

broken down by organ systems, based on physical structures. But when it comes to health and disease, organ systems are rarely affected in isolation. Distinctions like "heart health" and "brain health"—the sort still made on everything from cereal boxes to infomercials to ranking of academic medical centers—are outdated. So I divided this book not by traditional organ systems, but by categories of use. Most of the entries can be read in isolation but make the most sense in the context of the others, as read sequentially.

Overall, the book is predicated on something closer to the 1948 definition of health. It is drawn from my experience as a physician and journalist, and the people I've had the opportunity to meet throughout the course of my career so far, and whatever wisdom I gained from knowing them.

PART ONE

APPEARING

THE SUPERFICIAL PARTS

Butterfly children" are so called because their skin is like butterfly's wings. The name is meant to convey extreme fragility. But the weakness in butterfly wings is only a product of the fact that we are some one hundred thousand times larger than butterflies. In terms of biomechanics, these wings are actually paradigms of efficiency: light enough to be operated by a flying worm a fraction of their size, yet strong enough to hold up under the intense shear force of the wind and torrential rain that would be for us like standing under Niagara Falls.

The skin of a butterfly child, on the other hand, is rather an abject failure of biomechanics. Because of one detail. The formal name of the disease is dystrophic epidermolysis bullosa, or DEB. It's traditionally considered a pathology of the skin, the domain of the dermatologist, because it renders the skin like tissue paper that has been left in the sun. Such skin falls apart at the lightest touch. The condition has no cure. It is the worst disease you've never heard of. I write that without presuming which diseases you have heard of. The trademarked motto of the Dystrophic Epidermolysis Bullosa Research Association is "the worst disease you've never heard of." Its current executive director, Brett Kopelan, coined the phrase in earnest.

His daughter Rafi was born in a Manhattan hospital on November 19, 2007. Her mother, Jackie, was more than a little concerned that patches of skin were missing on their newborn's hands and feet. She had been two weeks past due, and the doctors initially reassured Jackie and Brett that their baby had been "overcooked." But the casual dismissal proved too casual when over the next few hours Rafi began bleeding. Nurses rushed her to the intensive care unit. There she would spend the first month of her life in complete isolation, undergoing a battery of tests, unable to be touched by her parents. After two weeks the doctors came to the Kopelans with a potential diagnosis, a name that would become their lives.

"They think it's something called *epi-dermo-lysis . . . bullosa*?" Brett recalls saying in a harried phone call with his brother, who is chief of surgery at a hospital just across the river in New Jersey, to which the surgeon replied, "Oh, shit." Brett ran to Google and read about DEB. His first thought was that it was the worst disease he'd never heard of.

On the short arm of the third of your twenty-three chromosomes sits a gene called *COL7A1*. It is responsible for the production of the protein that assembles collagen VII. Collagen proteins constitute all of the connective tissue in our bodies, and a third of our total protein. From the Greek for "glue," collagen holds together everything from skin to ligaments and tendons. It comes in several known types (of which collagen VII is one).

Epidermolysis bullosa is a rare disease in several ways, not least in that much of the problem traces to a discrete gene. Most diseases are far more complex than any single gene can explain.

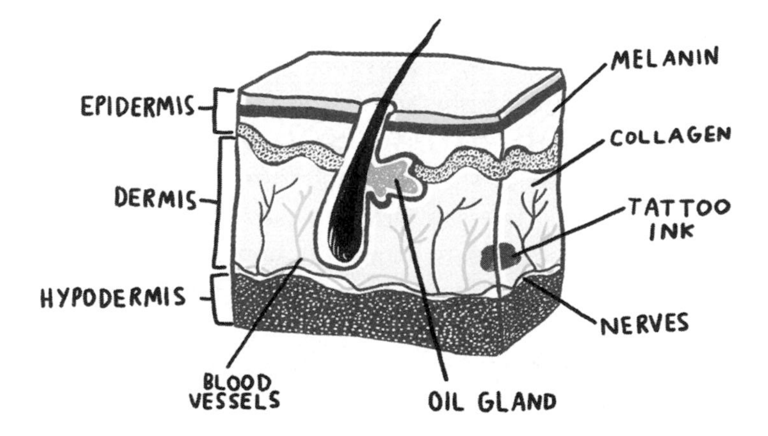

But mutations in the *COL7A1* gene seem to be responsible for all three major forms of dystrophic epidermolysis bullosa, of which Rafi's is the most severe.

Collagen VII anchors our outer layer of skin (epidermis) to our base layer (dermis). Without it, the layers separate and the skin crinkles and blisters, coming off at the slightest provocation. When Rafi reflexively scratches an itch, she wounds herself. The seams on her shirt cause blisters. Many mornings she wakes up with her pajamas pasted to her skin in multiple places by dried blood. The extrication is grueling.

And because collagen VII provides structure throughout the organs of her body, this affects not just the skin, but the internal organs as well. Blisters and scars within her mouth and esophagus make it difficult to chew and swallow food. She has eye inflammation that can lead to blindness. She has a very high risk of developing an aggressive type of skin cancer at a young age. She has osteoporosis, syndactyly (fusion of the fingers), and mild heart failure.

Rafi's form of epidermolysis bullosa affects fewer than one in a million infants. For those who survive, life does not involve much interaction with other people. So it is a disease of people whom we are not likely to come to know. The spectrum of what most of us consider normal in our day-to-day lives is skewed strongly away from conditions like DEB, and toward small blemishes. If it weren't, we might be more appreciative of the skin that we have, and the simple fact that it adheres to our bodies.

The average person has about six pounds of skin. Like most (though not all) organs, it's essential to life. If you woke up one day and your skin had vanished, you would quickly die. In what remained of your short life, there would be problems socially. It's the largest and most dynamic organ in the human body, constantly turning over and regenerating. Skin, along with hair, is unique among body parts in that it is dead cells we carry around. In any other organ, dead cells are discarded. But the cells in skin and hair stay along with us for a while and serve important functions, not least of which is social identity and thus the foundation on which the understanding of ourselves is built.

The skin we had last year—last season, even—is not the skin we have today. Most of the cells that compose our bodies are constantly dying and being replaced. Around 8 percent of our genes are not even human, but viral. We are born with viruses woven into our DNA, and we contain trillions of bacteria that are responsible for, among other things, the appearances of our faces, our body weights, and our states of mind. Our bodies are dynamic networks of genetic information shaped by experience, and microbes that change who we are in every moment. We are born with signals that will tell us to go bald when most people would appraise us more favorably if we had hair, and to be anxious when we needn't be, and to get cancers that we tried hard to avoid. The doling out of years and health and happiness will not be fair.

The seemingly superficial parts, and the way they are perceived by ourselves and others, accumulate into how we understand ourselves, and then into how we move through the world and treat one another.

How can I tell if I'm beautiful? I mean in the purely superficial physical way that I know I shouldn't care about but do because I am a person who exists in the world.

In 1909, Maksymilian Faktorowicz opened a beautification establishment in Los Angeles. Under the name Max Factor, he would become famous for his cosmetic products, which he sold as part of a pseudoscientific process of "diagnosing" abnormalities in people's (mostly women's) faces. He did this using a device he invented called the "beauty micrometer." An elaborate hood of metallic bands held in place by an array of adjustable screws, the micrometer could be placed over a woman's head and, as one of his ads at the time claimed, flaws almost invisible to the ordinary eye would become obvious. Then he could apply one of his "makeup" products, a term coined by Factor, to correct the flaw in this person: "If, for instance, the subject's nose is slightly crooked—so slightly, in fact, that it escapes ordinary observation—the flaw is promptly detected by the instrument, and corrective makeup is applied by an experienced operator." Even if putting on a metal hood that could tell people exactly why they're not beautiful didn't seem wrong on infinite levels, there was also the problem that Factor's micrometer was contingent on an empirical definition of beauty. A device that tells people what's wrong with them is predicated on an understanding of what is right. Max Factor's approach is a textbook example of the sales tactic that is still so successful in selling body-improving products: convince people that there is a deficit in some concrete way, and then sell the antidote.

In the case of facial symmetry, some evolutionary biologists do believe that we are attracted to symmetric faces because they might indicate health and thus reproductive viability. From a strict perspective of evolutionary biology, someone with a prominent growth spiraling out of the side of their eye, for instance, might be viewed as a "maladaptive" choice for a mate. Instincts warn that this person may not survive through the gestation and child-rearing process, possibly not even conception. Best to move on.

But today most people survive long enough to reproduce and care not just for children, but grandchildren, great-grandchildren, and even domesticated cats. We can be less calculating about who to mate with. We can and do afford ourselves attraction not to some standard of normalcy, but to novelty and anomaly.

While Factor was convincing everyone that they were empirically inadequate based on a standard of normalcy that he created to sell products, the University of Michigan sociologist Charles Horton Cooley proposed a more nuanced approach, called "the looking-glass self." The idea was that we understand ourselves based not on some empirical idea of what is right or wrong about us, but from how others react to us. It's difficult to believe you're physically attractive when the world treats you otherwise, and vice versa. "The thing that moves us to pride or shame," he wrote in 1922, "is not the mere mechanical reflection of ourselves, but an imputed sentiment, the imagined effect of this reflection upon another's mind."

Cooley repopularized the timeless idea that other people are not just part of our world, or even merely important to our understanding of ourselves: They are everything. Technically there are individual humans, just as technically coral is a collection of trillions of tiny sessile polyps, each as wide as the head of a pin. Alone in the sea, the polyps would be nothing. Together they are barrier reefs that sink ships.

The idea of a looking-glass self could seem disempower-

ing, in that our understandings of ourselves are subject to the perceptions of others. A less devastating way of thinking about a world of looking-glass people, I think, is the idea that everywhere we go, not only are we surrounded by mirrors, but we are mirrors ourselves. It's not the face in the Max Factor machine that matters, but the way that face is received. We can't always choose our mirrors, but we can choose the kind of mirrors we will be—a kind mirror, or a malevolent mirror, or anything in between.

Why do I have dimples?

The muscle that pulls the corners of your mouth up and back (a "smile") is called the zygomaticus. In people with dimples, that muscle is shorter than in the average person and may be forked into two ends, one of which is tethered to the dermis of the cheek, which then gets sucked inward when the person smiles. This is one way that beauty happens.

It's an anatomical anomaly, sometimes even referred to as a "defect." That understanding comes from an oft-cited theme in biology: that form necessarily correlates with function. Everything must happen for some reason, right? If dimples are a form without a clear function, then it's easy to dismiss them as defects. It would be easier to write this book if it were the case that our body parts either had a clear purpose or else represented *defects* or *diseases*. But we're more complex and interesting than that.

Biological function is a concept foundational to understanding health and disease, and it's defined most often as the reason that a structure or process came to exist in a system. We have opposable thumbs, by the etiological theory of biological function, because they gave us an advantage in using certain tools.

While form can help inform our understanding of function,

few cases are so clear-cut as thumbs. Some beards grow, some skin peels, and some cheeks dimple because they evolved to do so in certain people under certain conditions. There is no "purpose," teleologically, for beards to grow or skin to peel or cheeks to dimple.

Theoretically all functions should together contribute to our fitness—to keeping us alive and well, as populations—but they may not do so individually. In isolation, a particular bodily function—like sleeping, for instance—may seem to be only a weakness. Sleeping is a time when we might be eaten by birds. But it exists, according to leading theorists on the still outstanding question of why we sleep, because it augments the functions of other body parts.

Elements of our forms may also be vestigial, like wisdom teeth or appendices, relics that lost function over time as systems changed. There is a spectrum of vestigiality, with some parts trending toward obsolescence but not yet useless. Other parts may never have been functional, but simply emerged as side effects of the functions of other parts. (These are sometimes known as "spandrels," an allusion to decorative flourishes in architecture that serve no purpose in supporting the structure.)

The overarching idea is that almost no body part can be explained in isolation. They make sense only in the context of whole people, just like whole people make sense only in the context of whole populations. In a parallel world, dimples are anomalies we would attempt to prevent or correct. But in this time and place we choose to desire and envy them. Sometimes even to create them by force.

If I didn't have dimples, could I give them to myself?

In 1936, entrepreneur Isabella Gilbert of Rochester, New York, advertised a "dimple machine" that consisted of a "face-fitting spring carrying two tiny knobs which press into the cheeks." Over time, this force should produce "a fine set of dimples."

It didn't, though, because that is not how dimples work.

If only as an alternative to that particular form of hell, we may be fortunate that today a surgeon can perform a twenty-minute procedure to suture a cheek muscle known as the buccinator to the interior surface of the skin of a person's cheek, creating a dimple. It can all be done without ever puncturing the skin from the outside, coming through the inside of the cheek and cutting out a little bit of the cheek muscle, then running a suture through that muscle to the undersurface of the skin in the cheek. Pull it tight, and the skin will pucker. This is all done while the person is awake.

The puckering is exactly the sort of outcome that cosmetic surgeons spend years perfecting the art of suturing in order to avoid. So it took a truly heterodox thinker to invent the procedure. Based in Beverly Hills, plastic surgeon Gal Aharonov considers himself the father of the American trend in dimple surgeries. "It wasn't a trend before I started doing it," he told me. That's a phrase I rarely trust, but he does appear to have created the dimpling technique, about ten years ago.

"There were a couple people elsewhere in the world doing it when I started, but it wasn't really coming out well. I thought it looked awkward and weird," he told me. "So I figured out how to do it, and I put a little thing on my website, and the next thing I know, I got contacted by a bunch of news outlets."

In 2010, he appeared on CBS's daytime television show *The Doctors* ("Where MD meets TV") with his patient Felicia, who had decided to "upgrade her smile." In the episode, Aharonov gives her a handheld mirror and marks the places where she'd like her dimples. "It's fun to take these people and give them basically what they've always wanted," he says, tonally betraying that he was less than enamored of himself in that moment. A few minutes later he finishes the procedure. Felicia looks in the mirror and says, "Oh my God, I have dimples." And it's true, she does. She looks happy. Though it's hard to say for sure.

Today Aharonov advertises the procedure as safe and effective, even though there is, he acknowledges, "usually a period of time after the dimple creation surgery is done where the dimple is present even when you are not smiling," which could be unnerving. But for people who are jealous of people with dimples, I suppose it's reassuring to know that they're a lunchtime procedure away.

That and, of course, a couple thousand dollars. In Britain, where the procedure had a moment in the wake of Kate Middleton's dimpled ascension, it costs the equivalent of $1,200 to $2,500. Aharonov charges $4,000.

Of course, when there is expensive cosmetic surgery to be had, there is inexpensive cosmetic surgery to be found. On the other side of the planet, the surgeon Krishna Chaudhari of the Cosmetic Laser Surgery Center in Pune, India—where Bollywood films helped spawn dimpleplasty's demand—practices an alternative approach to the procedure, which he demonstrates on his YouTube channel. Even though it's still pretty straightforward as surgeries go, watching it happen there is a surreal experience that I recommend to no one. Chaudhari's video is a montage of still images taken over the course of the operation, first as eight-millimeter holes are punched all the way through a young man's cheeks, and then as a suture is run through the dermis to anchor it to the buccinator muscle. It doesn't help that the lighting makes it look like it's being done in a basement or a cave, or maybe the basement of a cave, and it's set to transcendental instrumental music that could pass for a deep cut from *Dark Side of the Moon*. (If you have to have surgery, and you want to watch a video of the procedure beforehand, ask your surgeon to recommend one before you venture too deeply into the world of Internet surgery videos.)

Many cosmetic surgeons performing dimpleplasty today

are doing so by a technique of their own invention. Abdul-Reda Lari, M.D., who practices in Kuwait, spurns the full-penetration approach. He invented a technique that has gained such acclaim that surgeons come all the way from India to learn from him.

"I used to put scissors in and split the muscles," Lari told me. "Now I tend not to do that. I put a knife inside the mouth and scratch the dermis inside the cheek in a vertical manner, and the bolsters [a shaping device Lari designed] keep it in position for up to two weeks. If she's complaining, I can remove it earlier."

He let the pronoun slip there. Almost all dimple clients in Kuwait are women. The same is true elsewhere.

Lari's technique is more complex than most. It involves not one but multiple sutures, and a bolster that must be tied into the inside of the cheek and left there for two weeks. Because using a single suture can leave the dimple looking like an unnatural pinpoint, he believes he gets much better results than others: a vertically oriented dimple that appears only when the person is smiling. His method isn't as popular because the procedure requires a follow-up appointment and a little more discomfort. He has done fewer than one hundred cases. Most people choose the simpler technique because they prefer instant gratification. Plus, Lari says, "I charge on the expensive side, $1,000 for both sides. It takes me two minutes to do it," he says, laughing.

"How much does it cost in the U.S.?," he asked me. He seemed a little disheartened when I told him.

Virginia cosmetic surgeon Morad Tavallali likens the anatomy of a dimple to that of cellulite, which is created by fat infiltration into the skin. There is some potential space that can be filled by fat that has no other place to go except into our skin. But there are fibrous bands within the dermis that resist expansion, so they appear as dimples. Tavallali can do procedures to eradicate this dimpling in a person's thighs, and he can do procedures to create it in a person's face.

Beauty is only ever about context.

Though they're easy enough for him to create, Tavallali has reservations about surgical dimples. "In some cases, a board-certified cosmetic plastic surgeon may invent a new procedure," he writes on his blog, detailing how a dimple procedure can be performed, but he offers a disclaimer: "It is a surgery that not many plastic surgeons perform, and it's cute but can be problematic! Been there, done that! I no longer perform it!"

Any procedure that's undertaken at cost and risk to patients purely for purposes of conforming to societal norms of beauty "can be problematic." So Tavallali is less likely referring to the massive cultural implications than to the fact that the surgery doesn't always come out looking great. Or at least that it doesn't wear well over time. The long-term appearance of synthetic dimples is unpredictable because they depend on scar tissue. Everyone scars differently. As a spokesperson for one British plastic surgery group said, "Designer dimples could become designer disasters within a matter of years."

A more lucid case against getting dimples comes from Beverly Hills surgeon Aharonov. A decade after ushering in the trend, the creator of dimple creation is contrite.

"There was a period where I was like, 'This is great. This is *my thing*,'" he told me. And indeed, other surgeons still contact him, wanting to learn. It's low risk, high profit, and high demand—Aharonov still gets, by his estimate, twenty to thirty requests for the procedure a day. But, like Tavallali, he has almost completely stopped doing the procedure. He's less than happy with the results. By his estimate, 90 percent of the cases came out well. In 10 percent the dimples were maybe asymmetric, one deeper than the other, or they were too deep and didn't flatten out quite right when the patient wasn't smiling. "To me," he said, "90 percent is just not high enough when it comes to messing with your face."

Aharonov delves into the existential questions of cosmetic surgery. Why do people want an anomaly? Why do people get

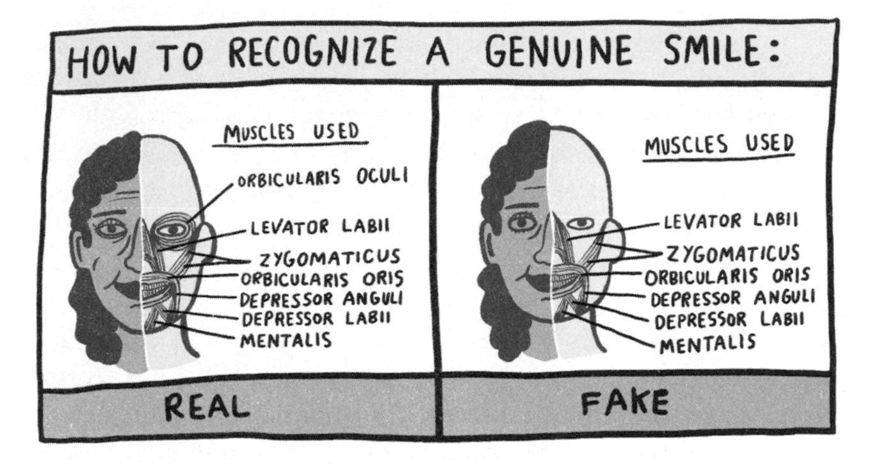

tattoos and piercings? "It's a desire to be different. A desire to be unique." Or the opposite: a desire to be like someone else they want to emulate.

In that way, these trends are not simply ridiculous, and these surgeries are not dumb. They are matters of social identity. But in the words of Spider-Man's uncle, with great power comes great responsibility. Cosmetic surgeons are arbitrators of motives. "I have to think, am I doing the right thing for this person?" Aharonov told me. "Do they want this surgery for the right reasons?"

Right reasons are difficult to articulate—perhaps undertaking the procedure for the sheer joy that the entire experience would bring to a person's life. But there are definitely wrong reasons. The number one rule of cosmetic surgery is not to aspire to perfection. Even the YouTube commenters—the most judgmental, barbaric people on earth and perhaps elsewhere—on Chaudhari's apocalyptic instructional video seem to empathize. As one described his takeaway, "It's painful and may damage your face but i mean if that's what y'all want I won't hate after all it's your body."

Why don't tattoos wear off?

On a sunny morning at my favorite coffee shop in Fort Greene in Brooklyn I met a woman who was covered in tattoos, and we chatted. She was working on a children's book about why people get tattoos. Some of her tattoos were on her eyelids. Every time she blinked and squinted into the sun, I got to read the words NO FEAR. And then all I could think about was the thought process behind that tattoo. It's the kind that she will see only when she's looking at herself in a mirror with one eye closed. The eyelid is the most painful place to be tattooed. Is that worth the money and pain? I will have to read her book.

Like plastic surgeons, serious tattoo artists discourage or refuse to leave their mark when they feel it is ill advised or hastily undertaken—especially if that tattoo is in a place as prominent as the neck or face. The philosophy is that tattoos should be undertaken for oneself, and not to impress or make a point to others. An eyelid tattoo walks that line. In this case, it tells everyone she meets about her. Such an ostentatious sentiment tells me that she probably has at least SOME FEAR, if not VERY MUCH FEAR. Why else go to such lengths to advertise fearlessness?

Tattoos also tell me that a person might have hepatitis. One of the most interesting statistics in virology is that people with tattoos are six times more likely to have hepatitis C. Which is not to say that tattooing *causes* hepatitis C, of course. (But sometimes tattooing causes hepatitis C.) Any needle passing through the skin can do it. Tattoo needles go through the epidermis, the outer part of the skin that flakes off, and into the dermis, which is rich in blood vessels, nerves, and, after a tattoo artist is done with it, globs of dye.

White blood cells recognize that dye as an interloper, a potential threat, and attack it. But the globs are too large to be cleared. Futile attempts account for the visible inflammation that makes fresh tattoos red for a few days, during which reasonable people wait to Instagram them. If it stays red for longer than a couple days, then you probably have yourself a good old-fashioned tattoo infection. Every couple years there is an outbreak of infected tattoos in the United States that's traced back to infected ink. Because it's being injected so deeply into the skin, it needs to be a sterile product, like the saline solution that a hospital would inject into your veins. That's why the Centers for Disease Control and Prevention recommend going to parlors that "can confirm that their inks have undergone a process that eliminates harmful microbial contaminants." There's no regulation of that standard, so how you define it is up to you. Some parlors will dilute their ink with tap water to save money, and you can ask them to promise not to do that. NO FEAR.

Whether the dye is sterile or not, your white blood cells will attack it. But they can't beat it. The dye globs are, as white blood cells would say, "too damn big." Eventually our immune systems just give up and resign themselves to living with these dermal intruders. Tattoos are about defiance and individuality, but also resignation.

How can I remove my tattoo?

It's technically illegal in many states to get a tattoo while intoxicated. One in five American adults now has a tattoo, and there is no strict research on what percentage of those people were sober during that process, but in my experience it's less than 100. And even sober people regret things sometimes—they change their allegiances, they fall in and out of love. The "golden rule" in tattooing, according to Fallen Ink Laser Tattoo Removal in Minneapolis, is never to get a name of a significant other, or anything at all "that is meant to symbolize your love for your significant other." Which is one way to go through life, I suppose.

The tattoo removal industry in the United States has grown 440 percent in the past decade. Spending on tattoo removal is expected to hit $83.2 million by 2018. It's not a bad business to get into. The level of skill required could be mastered in a weekend seminar by most chimpanzees. All you have to do is aim a laser at the tattoo and press a button. That laser breaks down the dye globs into smaller globs until they are digestible by white blood cells called macrophages in the skin and whisked away to tattoo heaven. (That is to say, excreted in the person's feces.) It usually requires multiple sessions and a few hundred dollars. Which is still enough to make getting a tattoo feel like a commitment of some sort.

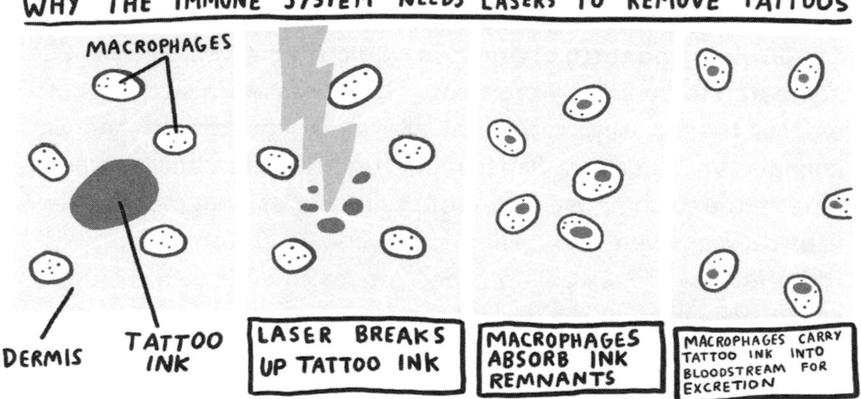

Can I get a more defined jaw by chewing gum?

This is actually a somewhat popular question among bodybuilders. At Bodybuilding.com, for instance, an anonymous twenty-five-year-old person asked, "Mah strong jaw brahs: Does chewing gum broaden the jawline/muscles?"

Though I don't generally endorse getting information from forums where people have their bench-press records appended to the bottom of everything they write, whenever you come across something addressed to "strong jaw brahs," it's worth a read. A fellow muscle-person in the forum goes on to throw in what sounds like misdirection: "I heard something about chewing on leather." But ultimately the bodybuilders who respond to the young jaw-seeker's question do provide a concise and erudite answer: "No one in the real world gives a **** about how defined your jaw is." (The asterisks are theirs, presumably in case any of the bodybuilders reading are children.)

At the Harvard University Department of Human Evolutionary Biology, Katherine Zink and Daniel Lieberman suggest that there is merit to the logic of the strong-jaw brahs. In the journal *Nature* in 2016, the researchers reported that our faces have been defined by our chewing habits over the course of centuries. The jawlines and teeth of previous *Homo* species were enormous compared to ours. They progressively decreased in size as *Homo erectus* began using tools that allowed them to hunt and eat animals, whose calorie-dense meat required less chewing. Once meat came to compose a third of their calories, that meant every year they chewed two million fewer chews. Combine that with the effect of stone tools on the "processing" of food—crudely chopping and grinding it—and the necessary force and stamina of our chewing apparati plummeted.

Then as now, when you don't use something, it leaves you. Many anthropologists believe that it is because we chew very little today relative to ancient humans that so many people need braces. Over generations, humans spent less time chewing as they were able to cook and farm; in the process, our jaws have slowly receded and shrunken, leaving our mouths crowded with teeth. Few of us have room for the third molars ("wisdom teeth") today. So they crowd in at an angle and push our other teeth into disarray. The need to prevent this by having the wisdom teeth extracted from our heads is a relatively recent phenomenon.

As smaller facial features began to arise, Zink and Lieberman argue, they may have actually been selected for as well. That is, our distant ancestors may have had preferences for smaller jaws. So it does seem that it is due to its relative rarity today, rather than any functional logic, that people seem to appreciate the appearance of Brad Pitt's face for his thorough jaw. Maroon 5's Adam Levine was almost certainly not named *People* magazine's "sexiest man alive" because of his musical prowess. Some have argued that Western attraction to angles probably comes from the association with a high-testosterone state, which signals virility and therefore reproductive viability.

If your mandible *did* continue to grow significantly as an adult, it would mean you had the serious hormone imbalance called acromegaly. This was the condition of the French actor and 1988 World Wrestling Federation champion André "the Giant" Roussimoff. His pituitary gland produced an abnormal amount of growth hormone as a child, and continued to produce growth hormone as only a child's should into adulthood, expanding him to more than seven feet and five hundred pounds. Even after the growth plates in his arms and legs closed, his facial bones continued to grow, giving him the bulky appearance of a storybook giant. Such characters were likely themselves modeled off of people who had acromegaly. Shrek, an "ogre," also had the classic structural features of someone whose life was defined by an excess of a natural, necessary hormone.

Not everyone with acromegaly becomes a giant; the subtler cases manifest as large hands, a large nose, and a prominent jaw. These effects are seen in athletes who take growth hormone as a performance enhancer, a potentially serious risk to continued existence. André the Giant grew until his heart could no longer support his body, the walls of his ventricles so thick and muscular that they could not be easily supplied with blood, and he died at forty-six.

Relative to taking growth hormone, then, chewing gum is benign. And possibly even beneficial to those who value being perceived as a strong-jaw brah. Our mandibles do tend to shrink throughout our lives, and that can be prevented. Just as osteoporosis can be stemmed with physical exercise, involution of the jawbone can be prevented by chewing often. (And the masseter muscle that goes around the mandible at the corners should, like any muscle, get at least a tiny bit bulkier with exercise.)

This is all most relevant as a reminder that we're adapted to eat high-fiber foods. With a concerted effort to chew gum or leather or foliage often, and to teach your children to do so, and to repeat that process for generations, you may eventually see a result.

But what about my chin? Can I make it more attractive?

We are the only hominids with true chins. If they evolved in the process of creating speech or chewing, they wouldn't be expected to vary much in size and shape between males and females. But they do. The evolutionary concept of sexual dimorphism explains that chins evolved as they did because of mating preferences. Fret not for the shallow superficiality of the day; we've been shallow for millennia.

In the case of chin shape, or the lack thereof, the term used among doctors is "submental fullness" ("sub" meaning below; "mental" being derived from *mentum,* meaning "chin" in Latin; "fullness" meaning fat). As the Harvard-trained dermatologist Omar Ibrahimi explained it when we spoke, "The submental area plagues a lot of men and women."

Ibrahimi practices in the affluent coastal city of Stamford, where he runs the Connecticut Skin Institute. He explained that submental fullness is an equal-opportunity focus of anxiety. "It doesn't just happen in overweight people," he told me. "As you age, you lose bone mass, and fat can collect in stubborn pockets."

The first step toward ameliorating submental fullness is the same as when trying to eliminate all bodily fullness: Eat well and move. (This is not part of the Hippocratic oath, but it might be added: *Make sure everyone is eating reasonably and being active, always, even if it means you sound pedantic and judgmental. Oh, yes, also: Do no harm.*) Still, a survey by the American Society for Dermatologic Surgery found that 68 percent of consumers are bothered by submental fullness, which is slightly higher than the number of Americans who are overweight or obese. So explained George Hruza, president of the society, in a press release for the company Kythera Biopharmaceuticals in the spring of 2015. He added an optimistic endorsement: "Kybella provides physicians with the first non-surgical treatment option to satisfy this unmet patient need."

That spring, the U.S. Food and Drug Administration approved Kybella for the "treatment" of submental fullness in humans. It's a drug meant to be injected into the neck, where it causes adipocytes (fat cells) to lyse (explode). And it's not surprising that it works. The sole ingredient in Kybella is deoxycholic acid, which is a bile salt—the exact same acid that is produced by the gallbladder and released after a meal to help the body break down fats in the small intestine.

Ibrahimi was among the first American physicians to begin using Kybella in 2015. "Is a double chin ruining your selfie photos?" reads the consumer-facing copy on the website of Ibrahimi's medical practice. In the lower corner of the page is a little animated GIF of the father of modern medicine, Hippocrates, spinning in his grave.* "Or do you work out and eat healthy but just cannot get rid of your double chin? Well, we have some very incredible news for you. The FDA just approved an injectable called Kybella that can eliminate a double chin with a series of quick injections."

When we first spoke, Ibrahimi was just about to fly to San Diego to be among the first 150 physicians trained by Kythera Biopharmaceuticals in how to use their product. Kybella is part

* Not really.

of a larger trend that he's seeing in cosmetic surgery. People are moving away from surgical procedures and toward injections, a trend that he believes is related to several celebrities who had complications after cosmetic procedures. In a relative way, injecting bile into people's chins may be a step toward reason.

The concept draws on a storied tradition of humans injecting body-shaping substances. The practice of mesotherapy started in the 1950s and had a moment in the 1990s, when people just kind of mixed vitamins and injected them into their anywhere, based on claims that had no basis in reality. Southern California and Brazil gained reputations as hotbeds of experimentation in "noninvasive body contouring." There were complications, and nothing worked especially well.

But dermatologist Adam Rotunda and biochemist Michael Kolodney at UCLA saw something in the concept. They became interested in creating a type of mesotherapy that had an actual scientific basis and could be proven safe. By 2005 they had filed a patent for using deoxycholic acid.

Unlike its precursors, the product had commercial appeal because it was "natural," in that bile acid is naturally produced by the body. The importance of that angle—here and in so many health messages and products—cannot be overemphasized when it comes to its marketing. (Even if there is really nothing natural about injecting bile acids into one's chin.)

Ten years later, the technique had undergone Phase III clinical trials and was FDA approved. The most common complications from Kybella are swelling, bruising, pain, and "areas of hardness," the FDA warns, if the acid leads to internal scarring. And because the acid destroys fat, the injection can damage nerves. (Nerves are coated in myelin, which contains fat.) The agency adds that this nerve injury can cause "an uneven smile or facial muscle weakness, and trouble swallowing." It costs around $1,500 per injection, and most people will need two to four injections before they see results.

But it's natural.

Why are some eyes blue?

Disassemble a blue human eye, and you will find nothing blue. The same with hazel or gray. All of our eyes contain the same pigment, a dark brown substance called melanin. It is the same pigment that gives skin and hair its color. We have this one pigment, which becomes an array of colors based on how and where it is concentrated.

The iris consists of two layers, the stroma in front and epithelium in back. Interplay between these layers results in a mix of absorbing and scattering incoming light and reflecting it in a way that produces eye color. It's a concept called structural coloration. The ultimate effect is produced only in the context of the entire eye as it exists.

What causes red eyes in photos?

Light reflects off the back of the eye, the retina, which is full of blood vessels. The retina is directly connected to your brain by way of the optic nerve. Some consider that nerve to be an extension of the brain. It's the closest most people get to photographing a friend's central nervous system.

What is a deviated septum?

Photographs of Eli Thompson spread over the Internet in 2015 after BuzzFeed published them under the title "Meet the Very Cute Baby Who Was Born Without a Nose."

{click}

The celebratory web page has more than a million views, according to a large red number on said page. The most popular comment posits that Eli "is already way too awesome to even care about our opinions." Eli's mother, Brandi, is coddling and kissing him in a photo under a caption in large typeface that says, "He's perfect the way he is."

Eli was perfect in the same sense that we all are perfect, but not in the sense that he could eat without suffocating. He spent the first five days of his life in the intensive care unit, where doctors made an incision through the front of his neck and his trachea, inserting a tube through which he would breathe for the remainder of his life. The air came in and out of his trachea below his vocal cords, so he made no noise when he cried. If he wished one day to speak, he would have to put a finger over the opening to produce a sound.

There is a possibility that a team of otolaryngologists and craniofacial plastic surgeons could construct a nose for Eli at some point, allowing him to dispense with the tracheostomy (from *stoma,* for "hole"). But it would be far from straightforward. Congenital arhinia, or being born without a nose, happens because of a "missed step" during formation of the embryo. With the construction of the nose comes the formation of nasal passages that connect the nostrils to the trachea. Because Eli's did not form, his brain sits lower in his head than do most people's. Attempt to create a nose for Eli in the typical place one expects to find it, and you may end up exposing his brain.

He is cute, though, it's true. His eyes are enormous, and he always seems to be smiling. His photos continued spreading around lifestyle blogs, even on the celebrity gossip blog of Perez Hilton, for nearly a year after his birth. Each blog post announced it as if it had just happened, celebrating how this very cute baby is doing just fine. Many were widely shared on social media. This treatment would not be afforded to most people with congenital deformities of the face. Baby Eli seems to allow people to indulge curiosity without upsetting sensibilities.

That combination is nearly as rare as the condition itself. Complete congenital arhinia has occurred in only around forty known people alive today. This is a wonder, given how intricate the process of nose formation is. The nose actually begins as two separate tubes that must merge in the midline to form one nose with two nostrils. In the fifth week after conception, two ridges called nasal placodes emerge from the face-to-be. These must quickly grow into what will be known as the medial and lateral nasal swellings, between which emerge negative spaces called nasal pits. By the end of the week, the medial swellings fuse to form the recognizable nasal septum. This will forever separate the nostrils (unless a person decides to put a hole through it with a large-bore needle for purposes of social identity). When this does not occur symmetrically, people are born with a "deviated septum" that can cause serious breathing impediments and/or snoring, which can render a person unlovable.

As the nasal pits deepen during the seventh week, the palate and nasal cavities emerge. In a case like Eli's, when there are no nasal pits, there is no palate or nasal cavities. By age one, our noses are 80 percent as wide as they will ever get. Our noses grow outward an average of 2.1 centimeters between age one and age eighteen.

No known cause of the arhinia like Eli's has ever been identified. A team of surgeons in China conducted a review of all known reports, concluding that while there may be some genetic predisposition, it is most likely the result of some aberrant signaling during the formation of the embryo. The surgeons recommend that even though the process of creating a nose can be tremendously complex—involving fracturing through the maxilla, creating nasal passageways and nostrils of cartilage that will require several stages of forcible expansion and long-term stenting—creating the nose is worthwhile whenever possible for purposes both physiological and psychological.

Why don't body hair and eyelashes keep growing, but head hair does?

The celebrity Elizabeth Taylor had at least one extra row of eyelashes on each eye (known as distichiasis), often the result of a mutation on a gene called *FOXc2*. In Taylor, people generally found it captivating. Like most genes, though, *FOXc2* doesn't affect just one bodily feature; it's involved with the development of the lungs, kidneys, heart, and lymphatic system—the lymph nodes and vessels that carry fluid and white blood cells to and from those nodes. People with an extra row of eyelashes may have a syndrome called lymphedema-distichiasis, in which the lymphatic system doesn't work properly, and the body retains fluid, and the heart can fail. In 2011, Taylor died of heart failure, which may or may not be related. It's easy to envy other people's eyelashes; it's not always the best use of our time.

Eyelashes do grow, they just fall out at a certain length. This topic was briefly addressed in a book by physician Beth Ann Ditkoff, *Why Don't Your Eyelashes Grow?* The hundred-plus similar questions in the book are the product of Ditkoff's young children, who remember to question what many of us take for granted about the oddness of human bodies. She explains that eyelashes simply fall out after about three months, unlike the hair on your head, which can grow for years without falling out.

Like all hair, lashes come from follicles, the smallest organs in the body. Hairs go through three phases. The length of all body hair depends on the length of the first phase, called anagen. When its time is up, anagen turns to catagen. The outer part of the root is cut off from its blood supply, and the hair stops growing.

After a couple weeks of catagen comes telogen, wherein the

follicle transitions into a resting state. For three months, then, the hair is called a "club hair." It is, like so many people in clubs, outwardly fine-looking but actually dead at its roots. It will either snap off or get displaced by new hair that rumbles up from below. For better or worse, each follicle is on its own time cycle, so we don't shed all our hair at once.

The real difference between head hair, arm hair, and eyelashes is the length of that anagen phase. On your head, it lasts a few years. Elsewhere it's more like a month. Were it otherwise, eyelashes and arm hair could grow to unwieldy lengths.

Rare outliers have very long anagen phases on their heads, so they can grow their hair to the floor. Others have very short phases, so they are not bald, but they also never really need a haircut. Stress can signal anagen to end prematurely and, in extreme cases, can lead to near-complete short-term loss of hair. But it generally grows back.

Eyelash "growth serums" can be found in the cosmetics aisle at drugstores and Walmart. They're usually just mixtures of peptides (parts of proteins), and they can cost a lot. One,

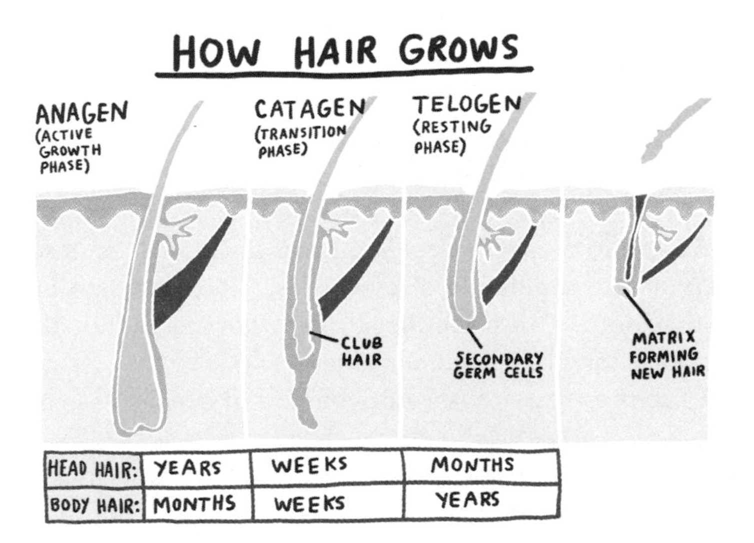

RevitaLash, is a proprietary blend of "natural botanicals" that *InStyle* magazine called "the Rolls-Royce of eyelash serums." I don't know what that means, but it costs $98 for two milliliters.

Prescription-grade eyelash serum is different, in that it actually makes eyelashes grow. It contains a small amount of the glaucoma medication bimatoprost, which came to market for eyelash enhancement after researchers noticed that people with glaucoma seemed to be developing more prominent lashes. It was serendipity in pharmacology, similar to the way Viagra was invented—when researchers testing it out as a blood pressure medication noticed an abundance of blood-engorged penises. Bimatoprost is sold under the product name Lumigan when it's used for glaucoma, and a more gender-targeted Latisse when it's used for eyelash enhancement.

After cataracts, glaucoma is the second leading cause of blindness worldwide. It is about seven times more common in black Americans than white Americans; but black Americans are less likely to be treated and twice as likely to develop visual impairment, often because of lack of access to health care or basic screening for glaucoma.

Meanwhile, some people pay handily for the same product to have better eyelashes.

Could I get rid of my eyelashes? If I'm tired of the eyelash game and I just want out?

In 2015 a group of mechanical engineers at Georgia Tech set out to determine the purpose of eyelashes. "Eyelashes are ubiquitous," they note in the scientific journal *Interface,* "although their function has long remained a mystery."

And so they tested the aerodynamics of eyelashes in a wind tunnel.

Mystery no more, the engineers found that eyelashes effectively protected their sets of model eyes from airborne debris and surface dehydration by a factor of two. "Short eyelashes create a stagnation zone above the ocular surface," the researchers reported, "causing shear stress to decrease with increasing eyelash length." However, longer eyelashes also channeled air toward the surface of the eye, causing shear stress to increase. These competing effects result in a minimum shear stress for people with medium-length eyelashes.

Which is to say, in eyelashes, as in all things, moderation. The eyelash enhancement industry is predicated on arbitrary beauty standards that it created. Prescription eyelash-growth serum may have some functional benefit for people who are genuinely deficient of lash and spend a good amount of time in wind. But generally, my advice is to avoid serums. The same goes for elixirs and tonics. If you see a potion, take a chance.

What makes hair curl?

Hair is made of the most abundant type of protein in the body, keratin. The traditional wisdom is that bonds form between sulfur molecules in the hair and cause the keratin filaments to kink and bend back on themselves. Hair straightening products break these bonds chemically or, in the case of hot flatteners, physically. Easy.

As with most things, the real explanation is more complex. In this case, fascinatingly so. Physicists at MIT recently set about creating a model of all the forces involved. In the physics journal *Physical Review Letters,* they explain their work, which I found hilarious for its length and tedium. A taste:

> We combine precision desktop experiments, numerics, and theoretical analysis to explore the equilibrium shapes set by the coupled effects of elasticity, natural curvature, nonlinear geometry, and gravity. A phase diagram is constructed in terms of the control

> parameters of the system, namely the dimensionless curvature and weight, where we identify three distinct regions: planar curls, localized helices, and global helices. We analyze the stability of planar configurations, and describe the localization of helical patterns for long rods, near their free end. The observed shapes and their associated phase boundaries are then rationalized based on the underlying physical ingredients.

Just reading that makes my hair curl! (Sometimes if you make a joke like that, people get distracted by laughter, and then you can change the subject so you don't have to actually pretend to understand.)

I got in touch with the lead researcher, Pedro Reis, an associate professor at MIT, to see if he could break down curling for me in a way that I might understand. He said that he could not. But he referred me to someone especially expert on the subject of hair curling, Basile Audoly at Institut d'Alembert in Paris. Audoly, too, deferred to a greater authority, Manuel Gamez-Garcia, who studied electrochemistry at the Tokyo Institute of Technology before taking a doctoral degree in engineering physics from Montreal University. For the past eighteen years he has devoted the entirety of his intellectual endeavors to the study of human hair, working for a company called Ashland that develops products for Procter & Gamble, Unilever, and L'Oréal. (This industry is where research like this originates, because who else would do it?)

Gamez-Garcia explained to me the details of what he had just presented at the TRI International Conference on Hair Science. (The world is bigger than we know.) When that became too much, he tried to simplify it in writing, which amounted to an email outlining eighteen ordinal points describing the anatomy of hair.

Basically he came to understand the behavior of hair by

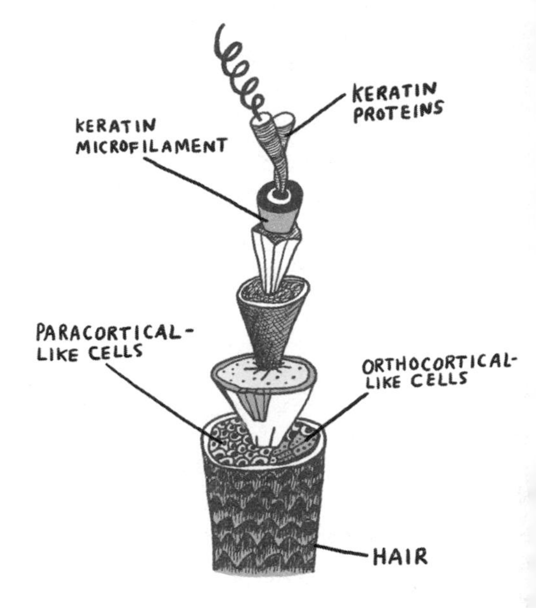

first understanding its structure. The active organs that are our hair follicles are constantly stacking threadlike microfilaments, made of keratin, onto one another. Together those microfilaments form a hair fiber. Each microfilament is tiny, but together they build a sturdy hair, meant to stand up against mechanical stresses of the environment. When the wind blows, your hair doesn't crack in half. (If it does, call someone.)

So all hair is fundamentally the same, but it ends up looking very different depending on how the internal filaments are arranged. Follicles use two main types of cells to arrange these filaments. The filaments in *paracortical* cells have a random mix of orientations: Some are parallel to the main axis of the hair, and others are at an angle. The filaments in *orthocortical* cells are all angled. Straight hair is mainly paracortical cells, and curly hair (depending on its degree of curliness) is about half orthocortical.

The way those filaments are stacked is not something that can be changed. Even when you pull or sleep on hair, or smash it into a flattening iron, the microfilaments will eventually stubbornly drive the recovery of a curl. Some things in nature are just not straight.

But that doesn't stop people from trying. On this curling, Gamez-Garcia has built a career. Fifteen years ago, he was devoted to meeting demand for the fashion of a permanent wave. "That has declined," he said. "People are looking now, for some reason, for straight hair. And changing curly hair to straight hair is much, much more difficult than vice versa."

The demand now is specifically for a "natural" straightening product—the Holy Grail for Gamez-Garcia and his competitors in hair product research and development. People were using formaldehyde for a while, but that raised safety concerns. The hair product companies for whom he works now want to promise something devoid of "harsh chemical techniques."

In essence, they want a natural way to undo the extraordinary complexity of nature.

When I shave or cut my hair, does it grow back faster?

The idea of being beaten and coming back stronger is inspiring, but not relevant to hair. When a young person breaks a bone, the fracture does tend to heal in a way that leaves the site stronger than it was before it broke. Muscle fibers grow back stronger when they are broken down. We might imagine, then, that the body's follicles react to shearing of the hair by pushing forth a handsome swath of thick, warm, protective hair—little follicles refusing to be silenced. But no. Like most parts of the body, injured or otherwise modified follicles do not get stronger. If anything, they get weaker and more vulnerable to repeat injury. Waxing, shaving, and even aggressively tight ponytails are more likely to damage follicles than to strengthen them.

Am I tall enough? If not, can I still get taller?

In 1981, a janitor at Dallas–Fort Worth International Airport noticed that he was getting taller. In the three years after graduating high school, the man, named Dennis Rodman, grew from the American average of five foot nine inches to the top percentile of humans: six foot seven inches. At that point he decided to try the sport of basketball again (he didn't make his high school team years earlier).

He quickly attained proficiency. Four years later, Rodman

would be selected in the second round of the NBA draft by an upstanding team called the Detroit Pistons, with whom he would win two national championships. Then he won three more with the Chicago Bulls, securing him a place in the NBA Hall of Fame.

Redemptive stories like this are used to console kids like me when we don't make the high school basketball team—to inspire us that anything is possible. Sometimes it is. Though it's obviously unlikely that any one of us is similar to Rodman, in terms of bone growth or otherwise. Because a radiologist could have looked at X-rays of the twenty-year-old janitor and said that something there was not *normal*. Could these really be the bones of someone so old?

By looking at X-ray images, a radiologist can determine the age of a child, based on the patterns of mineral deposits in the bones, their size and shape, and the amount of cartilage that has yet to become bone. These "bone age" X-rays are a common test in children's hospitals. If the child's calendar age is significantly different from the apparent age of the bones, this might be indicative of a hormone abnormality or malnourishment, and is sometimes an important indicator of child abuse. (It's unfortunately common for the signs of abuse to be identified, before anyone else, by a radiologist.)

One of the most important elements of determining the age of a young person's bones is the growth (or epiphyseal) plates. Found near the ends of our linear bones, these are the generators of new bone material that allow bones to grow longer throughout childhood and adolescence, while still being strong enough to support the walking, running, and jumping that young humans tend to enjoy. These almost always disappear between ages thirteen and eighteen. Coinciding with the end of a person's growth, these bands of bone-producing cells turn to bone themselves.

The odd thing about Rodman's X-rays would have been that

at age twenty his growth plates would have been clearly visible. Why would they stay open so long?

If you've ever squeezed an infant, you know that their bones are not bones, but cartilage. In the first few years of life, those cartilage cells ossify into bone. The exception is the growth plates. They are made of cartilage-generating cells called chondrocytes, which are only one step more differentiated than stem cells, which can become any type of cell. Growth hormone travels from our brains through our blood to these chondrocytes, signaling them to divide. As they do, they crank out cartilage that extends the bone and then ossifies into osteocytes (bone cells). At the end of puberty, our chondrocytes turn off. The cartilage cells in the growth plates turn into bone cells, never to return. Bones like the femur become one solid entity instead of two caps and a cylinder. After that point, it's impossible for them to grow.

There is no shortage of people who will tell you otherwise, though. My favorite is the Grow Taller Guru, or as he calls himself in his Internet videos, "the GTG," with a finger stabbing in at the viewer to emphasize each letter. His name is Lance Ward. He has multiple YouTube videos with hundreds of thousands of

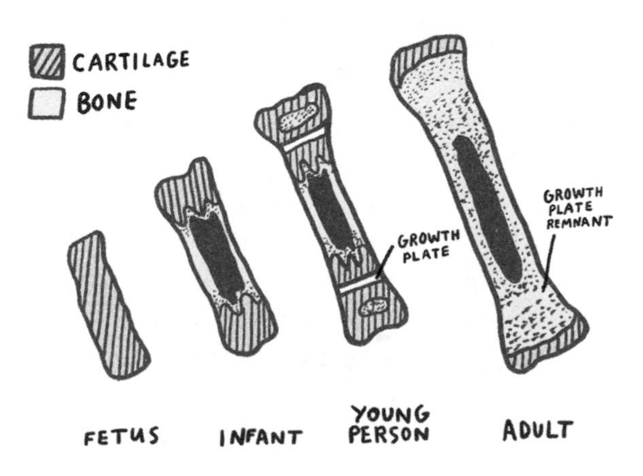

views, in which he promises that anyone can grow taller at any age. In "How Can I Possibly Grow Taller If My Growth Plates Are Closed?" he explains with signature exasperation, "Who's telling you about these growth plates? And that if they're closed, you might as well give up? It's like a virus. It's like cancer, man."

Ward urges viewers not to be passive agents in their destiny—not to accept the limits that society imposes on them, not to believe what they're told. He doesn't immediately explain how to grow taller, though. To access that information, you need to make a purchase. The purchase unlocks a secret method to which Ward is living testament. "I was just an average kid," he explains with canny forlornness in another video. "I wasn't particularly popular. All I wanted was a girlfriend." At age sixteen, he was five foot eight. He goes on to explain that he also wanted something more than a girlfriend—he became obsessed with the professional wrestler Goldberg, and decided he wanted to follow a similar career route. So he started looking for methods to grow taller. He bought some pills on the Internet. He bought some insoles that were supposed to stimulate the soles of his feet. He heard you were supposed to use them late at night, so he'd walk around in them at night. *It seemed like nothing was working*.

Finally, he started a secret regimen of vague bodily movements. By age eighteen, he was six foot two. To make things even more fascinating, his brother did the same thing, and it worked for him too, at right around the same age. His story is explained in a video called "How to Grow 3–6 Inches Taller in 90 Days." It is thirteen minutes long, and at no point does it explain how to grow any number of inches in any number of days. Yet the last I checked, it had 423,352 views on YouTube, which seems to indicate that the idea resonates with people.

To get the details of the secret height-inducing movement techniques that can make anyone grow taller at any age, you must first visit Ward's website, GrowTaller4U.com. I did. It offered in bold red text: "WARNING!!! You Will Get Noticed and attract Lots of Attention. . . . Being Tall Will Get You Instant Respect. . . . Having Increased Height makes you more Attractive and Desirable." It goes on like that for longer than I imagined any single web page could go, and to scroll down is to fall deeper into despair at the idea of people buying it.

Which they might, because those quotes about taller people commanding more respect, and generally being seen as more attractive, are demonstrably not false. And his math is hard to argue with: "As you will see from the DVD, you can see results of Half an Inch in just 7 Days! Which is an Inch in 2 weeks, 2 Inches a Month and 6 Inches in 90 days!"

The DVD is $97.03, plus $15.97 shipping. (Even for the purposes of journalism, I couldn't justify that shipping cost.)

If your growth plates are closed, the potential that you will get taller is precisely zero. For the benefit of everyone distressed by matters of stature, and for people living with congenital anomalies in limb length that make life difficult, I hope that one day bones can be easily lengthened after growth plates have fused. I do not believe the GTG will be the one to make it so.

It is true that X-rays of world-class athletes have shown that extreme exercise regimens can change our bones in adulthood to some degree. In their dominant arms, professional baseball pitchers and tennis players have both asymmetric muscles and thicker and longer bones. The difference is clearly visible in X-rays, though still minute (on the order of centimeters). The most relevant point is not about bone growth so much as bone maintenance; as with chewing leather to maintain the integrity of one's mandible, exercise keeps our bones strong.

But short of professional-level athletic training—and the tremendous wear on our joints that comes with it—our adult bones will not become longer or thicker. Doping with growth hormone and testosterone can add muscle for those bones to carry, but our chondrocytes ensure that skeletally mature major-league baseball players can control their bone growth only in width, not in length.

Though control over our own height is limited, people can and do influence the height of other people. According to researcher Daniel Schwekendiek at Seoul's Sungkyunkwan University, South Korean men average between 1.2 and 3.1 inches taller than North Korean men. Others put the number as high as 6 inches. When Dennis Rodman visited North Korea in 2013 and 2014 as a "basketball diplomat," he towered over them like a gender-fluid, heavily pierced Gandalf.

Schwekendiek explains that the difference in heights between North and South Koreans can't be because of genetics in any traditional sense. Korea was a single country until 1948, when the United States occupied the South and the Soviet Union occupied the North. The subsequent North Korean regime forced its people into poverty and malnutrition. Those who are not imprisoned by the state in gulags subsist largely on white rice produced in government-run farms, in a mode of laboring that appears only somewhat closer to freedom than imprisonment. The country does not trade with others, so the food supply is limited to what can be grown locally. There are almost no fruits and vegetables, and the crop yields are often poor. The agricultural industry is run by the federal government (as opposed to in the United States, where the largest agricultural sectors *run* the federal government). Not surprisingly, Schwekendiek has to base his measurements of North Koreans on refugees who were able to escape.

As with comparing one side of a tennis player to the other side of the same player, it's rare to have such genetically similar populations exposed to such disparate conditions for their entire lives. For scientists, data so free of confounding variables can induce a sort of intellectual orgasm. Of course, in the case of North Korea, where people live under a despotic regime laden with human rights abuses, it is a dour and guilt-laden orgasm.

The food-height relationship also gives a clear window into the role of lifestyle and environment that those of us living more comfortably regard as out of our control. Height is no exception from the dictum that we are far from beholden to genetics. Good food and exercise cannot make an impoverished North Korean child into Dennis Rodman. But it can add inches to a child who would otherwise be deprived. And if access to these basic needs is so intrinsic to our brains that it can trigger growth hormone to turn our growth plates on and off, what other elements of bodily health are similarly malleable?

It may be because we've long had an innate sense of height as a proxy for health that most people find taller people attractive, and so afford them all the luxuries that attractive people receive. ("With this person, I can have offspring who will survive," thinks our brain without telling us, only directing our sex organs to pulsate.) Sexual selection compounds the effects of natural selection that would favor larger stature—more advantageous for hunting and fighting, on the whole—making humans ever taller.

Thinking about height as a proxy for health is unfortunately not an outdated concept. The World Food Programme estimates that one in four children is chronically malnourished to the point of being stunted in height. If you know someone who is considering buying a $97.03 DVD, hide their money from them and either give it to the public schools to advance science education or use it to support the World Food Programme.

It's also critical to note that even though no increase in nutrition will make an adult person taller, overeating can make us shorter. Wander into any hospital and look at an X-ray of the human spine from the side, and it's supposed to be shaped like an S, curving forward in the abdominal region. Carrying extra weight in that area can pull the lower spine farther forward. Over time, it can also compress the discs between the vertebrae in the spine—which are largely water. When the American

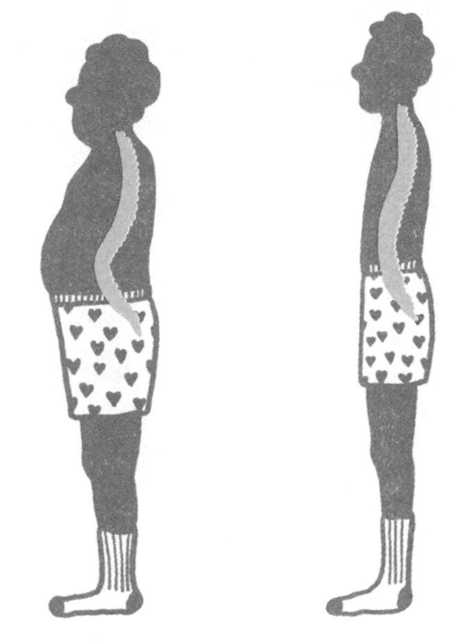

astronaut Scott Kelly touched down in Kazakhstan in 2016 after a year in space, he was two inches taller than his identical twin brother who had not been in space. Without the compression induced by gravity, the discs between Captain Kelly's vertebrae had expanded. This is what can happen when the effect of body weight is removed.

Gravity and aging are inescapable, but surplus body mass is not. Losing weight and building core muscles that support posture can allow the spine to return to its natural form, which leaves a person taller. Over time, too, this minimizes shrinkage by keeping bones and joint spaces from eroding and vertebrae from developing compression fractures.

I went to a boxing class recently where we warmed up with sit-ups, and the instructor said, "If you don't have a strong core, you don't have shit." I thought about that for a while. Ultimately, I disagree. But stretching and generally moving in ways that keep core muscles from fading to nothing does help maintain posture, and so minimizes stress on the spine and the discs between, which keeps them from desiccating and shrinking.

Please send me $97.03?

What are sunburns?

The World Health Organization classifies sunlight as a carcinogen. This might seem like an odd distinction for the thing on which all life depends. It's emblematic of the way we might think of all things that go into and onto our bodies.

When the sun burns out, there will be no more life on Earth.

Partly because before the sun burns out, it will expand and become so hot that it kills even the microbes that spill out of our desiccated corpses. Even now, at a comfortable distance and behind an ozone layer that remains capable of filtering much of the sun's harmful radiation, too much sunlight will indeed kill us.

But we don't see a lot of people marching and demonstrating against the sun. You don't see people demanding that their food be made sunlight free. Even though the sun will give millions of people cancer this year, most people have the sense that this would be absurd.

The sunlight that reaches Earth's surface comes in two main types of ultraviolet radiation: A and B. *Actually the spectrum of ultraviolet radiation is continuous,* says an objector somewhere, snorting milk up through his pharynx and out his nose. Okay, yes, A and B are how the wavelengths in sunlight are traditionally broken down in textbooks, but they're the same fundamental thing: energy that can be harmful to the skin.

Classically, UV-B was considered the "bad" kind of radiation most linked to sunburns and skin cancer, but later research pinned that on A as well. A sunscreen protects against A and B rays only if it's labeled "broad spectrum." Radiation breaks and tangles RNA and DNA within cells, which can of course lead to cancer. More often, skin cells are able to expunge damaged nucleic acids. That process triggers inflammation, which we call a sunburn. What we experience as a burn is actually the body protecting itself from cancer.

Ugly as that process is, human bodies require sunlight in order to function. Without it, the muscles weaken and the bones bend. It is only by exposure to the sun that our skin is able to produce the prehormone (a substance that is one conversion step away from becoming a hormone) known as vitamin D.

Further muddying feelings about sunlight, UV radiation is actually used to treat some skin disorders, like psoriasis. Patients receive "phototherapy." In other words, in the right context, a carcinogen is therapy. I saw it in action at the dermatology clinic at Vanderbilt University, which is part of a system of a network of medical clinics that occupy an old indoor shopping mall in Nashville. (You can walk around from the Vanderbilt gastroenterology clinic to the Vanderbilt neurology clinic to the Petsmart and Burlington Coat Factory.) In the Vanderbilt dermatology clinic, the phototherapy booth looks exactly like a tanning booth—and it is, just one that emits a narrow type of UV-B ray only.

Most skin cancers originate when radiation reaches the base layer of epidermal skin cells, causing their DNA to mutate. These cells are at least partly protected by the layer above, where melanin lives. The dark pigment is extremely effective at absorbing and dissipating UV radiation. And the darker the skin, the more protective melanin there is. It prevents not just cancer, but sunburns as well.

Melanin is produced when skin tans, too—a process of rapid adaptation of a body to an environment—which is why an already tan person is less likely to burn than they might have been at the beginning of summer.

And though the protective pigment melanin is an elegant solution to the problem of the sun—and a novel way of coloring our hair and eyes to make ourselves beautiful—I don't believe it's a stretch to suggest that it has also been at the core of more violence than any other single molecule in history. Blue-eyed women three hundred years ago were considered witches and burned at the stake. Of course now we realize the error of our ways. (People of any eye color can be witches.)

But in concert with the shapes of our faces and the curl of our hair, melanin still consistently underlies social divides of the sort that create and perpetuate more health problems than any textbook disease process.

. . . Totally. Wait, how?

On an April day in 2003, the Los Angeles City Council voted unanimously to create a new neighborhood, giving it the unadorned name "South Los Angeles." It was to start at Washington Boulevard, just south of downtown.

For people who knew Los Angeles, this might have read as a mistake. There was already a neighborhood there, called South Central. But after several decades of concentrated poverty and disproportionately high murder rates, the city believed South Central needed rebranding. So the neighborhood ten miles directly east of LAX airport and south of Beverly Hills is now South Los Angeles.

Rolled into it is the infamous neighborhood of Watts. The area continues to suffer from the extreme poverty and crime that plagued South Central. Watts has seen little fluctuation in affluence or safety, and has been untouched by gentrification seen elsewhere in the city. South L.A.'s population is now about 60 percent Hispanic and 40 percent black. It looks much the same today as it does in photographs from the 1960s, when the *Los Angeles Times* unabashedly referred to Watts as a "Negro district." Most of the 1.5 million people in South L.A.—the same population as Philadelphia—live below the federal poverty line. It is the largest contiguous area of poverty in the country.

It was in Watts on a hot summer evening in 1965 that a thirty-one-year-old white California highway patrolman named Lee Minikus pulled over a black man named Marquette Frye, reportedly on suspicion of intoxication. A bystander alerted Frye's mother, Rena, that Marquette had been pulled over, and she came running out of her nearby kitchen. By the account of Minikus, Rena encouraged her son to resist arrest. Punches were thrown. By the best accounts, the first came from Marquette. Minikus, who said in 2005 that if he could go back in time he would have done nothing differently, says that he beat the Fryes with his baton and arrested them. A crowd gathered at the scene. People booed as Rena was handcuffed and taken into custody. Someone broke a window, and then someone broke another. And then cars were burning, and stores and homes.

Eventually a thousand people were killed or injured and six hundred buildings were damaged or destroyed. Fourteen thousand National Guardspeople were deployed, and a curfew zone forty-five miles wide was implemented. It was a week before buses began running again and telephone service could be restored (the equivalent of everyone's cells and Internet today being down for *a week*—can you imagine?). By the end, more than thirty-five hundred people had been arrested.

In an attempt to deconstruct exactly how all of this happened, Governor Pat Brown commissioned an investigation by no less than the director of the CIA, John McCone. This was during the height of the Cold War—the end of the world seemed nigh, so McCone's plate was relatively full. The politically expedient thing for the McCone Commission to do would have been to blame one insubordinate poor black family and a landslide of mob psychology, as much of the news media had presented the situation. Many officials insisted that the Watts riots were the result of thousands upon thousands of vandals and miscreants. As though people suddenly decided to set fire to their community without cause.

A less reductive explanation is that this is the behavior of people with no recourse but to burn it all down. That is what the seventy members of the McCone Commission determined, after a hundred days spent embedded in Watts interviewing residents and scrutinizing citywide conditions. The causes of the Watts riots began long before Lee Minikus pulled over Marquette Frye. The causes of the Watts riots were poverty, inequality, and racial discrimination. The commission prescribed "emergency" literacy and preschool programs, job training, better low-income housing, public transportation, and, critically, health care access.

McCone specifically named the inciting factor in setting off these long-standing issues as a notorious November 1964 state ballot initiative called Proposition 14. California in the 1960s saw no shortage of social activism, but the hotly protested Proposition 14 stands out. The year prior, in June 1963, the state had passed the Rumford Fair Housing Act, which banned discrimination in the sale or rental of housing. That included lending institutions, mortgage holders, and real estate agents, with the implicit goal of equal opportunity for black home buyers. It seemed like a significant step forward for civil rights. It was also a controversial one. Even in Berkeley—*Berkeley*—a fair housing law had been narrowly voted down earlier that year. The following spring, though, the ballot initiative Proposition 14 was proposed to overturn the Rumford Fair Housing Act. And despite protests (which were matched by protests of the Fair Housing Act), it passed by popular vote. For the people living in poverty in South Los Angeles and aspiring to escape, or to procure loans to build and improve within their community, it was paralyzing. Even if the people of Watts somehow managed to make enough money to afford a place in Bel Air, they could legally be prevented from buying a home there. (The idea of a black family living in Bel Air was still novel when Uncle Phil and Aunt Viv managed it in 1990.)

Proposition 14 was eventually ruled unconstitutional by the U.S. Supreme Court. But in 1964, its passage was a clear sign to black Americans that not only was the system rigged—which they long knew—but a majority of people would explicitly vote to keep it that way. This was not just subtle inaction by the white majority, or the simple turning of a blind eye to the plight of the disenfranchised, but active oppression.

The events surrounding the arrest of Frye by Minikus were a spark inside scaffolding erected over decades, with fuel to rage a week and smolder indefinitely. The Watts riots were the point where government policy came into direct contact with the consequences of poverty and inequality. This would happen again when in 1992 Rodney King was beaten by officers of the law who went unpunished, and again in 2015 in Ferguson, Missouri, after the young man Michael Brown was shot repeatedly in the back by police officer Darren Wilson.

Among the dire conditions that led to the 1965 riots was lack of access to health care. It has become a common refrain in public health that a person's zip code is a better predictor of their health than their genetic code. The region of Westmont in South L.A. has a life expectancy ten years lower than across town in Culver City. Throughout South Los Angeles, one in three adults is uninsured.

This is all the work of people, not melanin, of course. Were we all identically pigmented, we would have found other ways to divide ourselves. The Watts riots are a classic example of systemic injustice, but the role of health care access in feeding the disparities in South Central is often left out of the story. Watts is a severe example of the gaps in the health equity that pervade the United States (and many other countries)—long before and after the neighborhood burned.

While much of the health care system has ignored these communities, a few practitioners have not. In South Central in July 1964, the year before Watts, in the back building of St. John's Episcopal Cathedral on Adams Boulevard, a small group of health care professionals began providing free care to the growing number of poor children. Volunteering their Saturdays, local doctors and nurses formed what came to be known as the St. John's clinic. It would grow into a staple of health care in South Los Angeles and a model for the country, explicitly serving those disenfranchised by race.

And, most recently, gender.

Why don't most females have Adam's apples?

In one of the early scenes in the books the Bible and the Torah, a male named Adam eats a "forbidden fruit." That fruit may or may not have been an apple. The books don't specify "apple," but that's how Byzantine artists chose to illustrate the scene. This apple got caught in this fellow Adam's throat—because, you'll recall, it was forbidden, and at this point in the books the antagonist, God, was not one for empty threats.

The apple stayed lodged in Adam's throat forever, somehow inexorably integrated into the tissue of his larynx. That became the basis for the term "Adam's apple"—what anatomists would call the laryngeal prominence.

I'm no scholar of religion, but I believe that Eve ate the apple, too? That was maybe an integral part of the story? Even as myth, the explanation has holes. It's also lacking physiologically. If a person had a large chunk of apple lodged in his larynx, he would retch and cough until it came out, because that is what the choking reflex is for. Indeed, if one is looking for evidence of the hand of God in the mechanisms of the human body, look no further than the gag reflex. The body attempts to save itself from peril by ejecting food without wasting time consulting a person's consciousness. (Only two-thirds of people have a gag reflex, so this answer is not meant to exclude people who do not, but is a simple assumption that this person's neural circuitry is in this regard consistent with that of the majority of the human population.)

When a nerve that goes to your tongue and throat—called the glossopharyngeal nerve (*glosso* = tongue, *pharyngeal* = pharyngeal)—senses something that is too large to pass through your throat, it sends a signal directly to your brain stem, which deploys a signal to contract the muscles in the pharynx. If Adam had no gag reflex and was also unable to cough because of some unique paralysis of these nerves—or if he had a stroke that short-circuited part of his brain stem—the apple might indeed have sat in his throat for a prolonged period. He would become known as wheezy Adam, or simply Adam the guy with the very high-pitched voice. Before long, the apple would begin to rot, and the area around it would become infected. Pus would gather until Adam's throat filled up and closed off entirely. He would die either of a mechanical suffocation or septic infection due to the fetid apple lodged in his throat.

The Adam's apple is not a structure that's unique to the male larynx. It's just one that tends to be less prominent in females. It is the cartilage that sits just above the thyroid gland, aptly called the thyroid cartilage. During puberty in males, testosterone stimulates growth of that cartilage, and with it the entire "voice box." That growth lengthens the vocal folds (or vocal cords). Like longer strings on musical instruments, the vibrations of longer vocal cords produce deeper sounds. Psychologists have shown that deeper-voiced males tend to be more attractive to mates, which seems to account for sexual selection in favor of a larger Adam's apple.

The trait gets perpetuated not because a big laryngeal prominence—a pretty useless structure—confers a survival advantage, but because big Adam's apples (and the deep voices that come with them) are attractive for what they allude to. Forming an Adam's apple requires testosterone, and that usually means functioning testicles. When we look up at a billboard in Times

Square featuring a man with a large Adam's apple, we are essentially being shown a demonstration of testicular aptitude. "Buy this product," the ad is saying, "and you'll have good testicles."

Because our larynxes are made of cartilage (and not bone), the Adam's apple can, like a person's ears and nose, continue to grow even after puberty. When a professional baseball player starts taking testosterone, his larynx can grow, an extension of the effect that happens during puberty, when testosterone levels in males increase several hundredfold.

This is the point when males and females become dramatically different in many ways that might seem unrelated to sexual maturity. At ten years of age, males and females can run at essentially the same speed. By the end of puberty, the top male runners are dominant over the top females. The average male can jump higher and throw farther than the average female by three times.

Part of the discrepancy is cultural—the result of more strongly encouraging athleticism in males—but even in elite athletes of both sexes who train from childhood, the differences between sexes persist. Males have testosterone levels that are two hundred times those of females, which seems to account for broader shoulders, longer limbs, and larger hearts and lungs relative to their bodies.

This was not always the case. Long ago, male and female humans were more physically similar. But the mechanics of reproduction have changed our appearances over time. Because human gestation takes nine months, one male could (historically) mate with many females in rapid succession, while females could not do the same. So there was then, as now, an excess of males. They had to fight with one another for opportunities to mate with females, leaving males with the traits that today we equate with athleticism.

This was only exacerbated by sexual selection, wherein females generally came to prefer mates who looked more discernibly male, even if the male was not technically more fit. And vice versa. This generally meant the appearance of a high-testosterone male was deemed preferable. In his popular book *The 4-Hour Body,* the author and "bodyhacker" Tim Ferriss recounts his own experiments manipulating his body's testosterone levels by eating enormous amounts of meat. As he describes it, women seemed somehow able to sense the testosterone emanating from him, and he became irresistible to them.

. . . Does that work? **Aren't we attracted by pheromones?**

The concept of pheromones—chemicals that we emit that make other humans want to have sex with us—is *deeply* intriguing. People have hypothesized that pheromones are by-products of testosterone and estrogen. Thousands of volatile compounds do emanate from our skin and breath, and from everything that comes out of us. But human pheromones haven't borne out scientifically. Ferriss might have been imagining everything. Maybe the women he describes were just attracted to his Adam's apple. Maybe there were no women at all.

The more socially pointed use of testosterone today is not in major-league baseball or sex gamesmanship, but for people who were born females and are transitioning from the physical carriage of a woman to that of a man. In transgender health, the use of the sex hormones (testosterone and estrogen) to affirm a person's sense of gender identity has recently gained status as a matter of medical importance in the eyes of most professional bodies of physicians, including the American College of Physicians, the American Medical Association, and the American Psychological Association. The U.S. Supreme Court has ruled that insurance companies cannot deny coverage of hormone prescriptions. As of January 2016, all U.S. federal employees are eligible for at least some forms of gender transition therapy.

This tremendous change in the status of health care for a long-marginalized population raises critical questions about justice and human rights in all health care. At least seventy-five countries have active laws criminalizing sex that is not between a male man and a female woman. Elsewhere the lines of acceptability are drawn in different places, and the discrimination against people who don't conform to traditional gender roles tends to be more insidious.

In the United States, rates of suicide among transgender people are estimated to be nineteen times as high as those among the remaining population. Though most of us are not overtly violent toward one another, we perpetuate notions that divide. In the realm of health care, the system is built almost exclusively around traditional concepts of gender.

Despite legal mandates and medical expert recommendation, gender-transitioning cases are consistently, routinely denied by both insurance companies and Medicaid. There are currently few places where the uninsured can get access to transgender health services—even insured people have a small number to choose from that are culturally competent and trained in health care for people outside of the gender binary. This is because most medical schools and residencies provide little to no education on the subject. There is no certifying or accrediting process. Historically, most care has been done on the black market in danger, or in reluctant medical settings where patients experienced discrimination, even hostility, from their health providers. But improvements to this situation have come from an unlikely source.

After the Watts riots, the dictum from the McCone Commission—that access to health care is fundamental to a functioning society—was largely ignored or forgotten. But the need for it was in evidence in the back room of St. John's Episcopal Cathedral in 1965. This makeshift facility grew into a pillar of community health in South L.A. By the 1990s, St. John's was a small but thriving clinic, known as a place that uninsured people could rely on.

That's when Jim Mangia came south from San Francisco, where he had been working at the peak of the panicked response to the AIDS epidemic, after finishing a public health degree at Columbia University. He's a few inches shorter than I am but feels taller, with a strong accent and the demeanor of a man still not adapted to California. He grew up in Brooklyn and moved to the Silver Lake neighborhood of Los Angeles when it was what he calls "ghetto." Over the past two decades, Mangia led St. John's as it became the largest community health network in South Los Angeles, serving seventy-five thousand patients per year. One clinic became fourteen. That accounts for about 40 percent of the primary care in all of South Los Angeles.

When Americans are uninsured, they are ultimately cared for by tax-subsidized programs. St. John's is a network of federally qualified health centers (FQHCs), which means that it's a nonprofit clinic that provides care to underserved and disadvantaged populations—those places where many people are uninsured. FQHCs receive tax credits and enhanced Medicaid reimbursement, and are eligible for grant funding. St. John's patient population includes many migrant and seasonal farmworkers, homeless people, and residents of public housing.

The FQHC program was established by Lyndon Johnson in 1965, one year after he had declared war on poverty, and the same year as Watts. In his 1964 State of the Union address, Johnson had said the program would "emphasize this cooperative approach to help that one-fifth of all American families with incomes too small to even meet their basic needs."

The approach Johnson described was one based on improving systems. "Our chief weapons in a more pinpointed attack," he said, "will be better schools, and better health, and better homes, and better training, and better job opportunities to help more Americans, especially young Americans, escape from

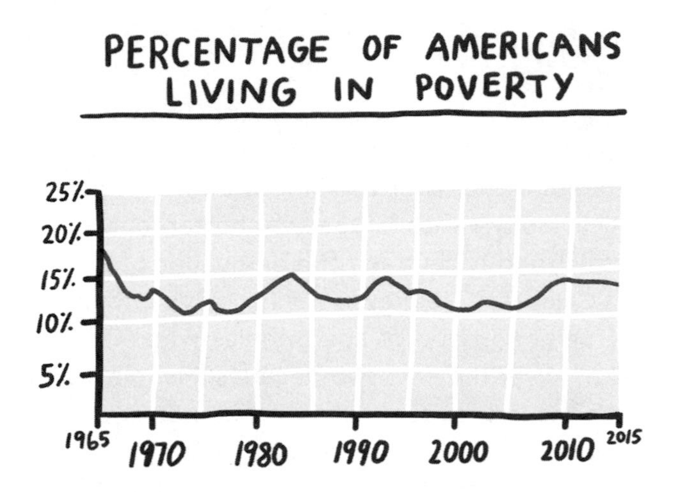

squalor and misery and unemployment rolls where other citizens help to carry them."

Like so many metaphorical wars to follow, the War on Poverty was unsuccessful; the one-fifth it aimed to address could no better meet their basic needs in 2015 than fifty years prior. It might be more accurate to say that the War on Poverty, as it was originally conceived, remains to be waged.

In the aftermath of Watts, with grocery stores destroyed and restaurants abandoned, the U.S. Department of Agriculture had to bring in ten tons of supplies. The monetary cost of this aid was in addition to the cost of enacting the National Guard, the damage to property (estimated at $100 million then, closer to a billion dollars today), and the judicial processing of thousands of cases and subsequent imprisonment. Johnson's idea was to have invested that up front, before the riots. The cost of a society as economically divided as the United States will be paid either way.

"Very often, a lack of jobs and money is not the cause of poverty, but a symptom," Johnson had said before the riots. "The cause may lie deeper, in our failure to give our fellow citizens a fair chance to develop their own capacities, in a lack of education and training, in a lack of medical care and housing, in a lack of decent communities in which to live and bring up their children."

Instead of targeting these ills more broadly, the central means of addressing poverty in the decade that followed became divisive welfare programs. *Government assistance* became the epithet with which all federal programs were branded, further dividing Americans along party lines. Democrats attempted to restructure systems to allow equality of opportunity, while Republicans saw those measures as unfair to people who had opportunities. The rift perpetuates itself today.

Yet one thing the War on Poverty does have to show for it today is a nationwide network of more than twelve hundred federally qualified health centers like St. John's in South L.A. The federal Office of Management and Budget has consistently rated the FQHCs among the most efficient federal programs. George W. Bush's approach to protecting uninsured Americans, rather than extending insurance to more citizens, was to pour money into more FQHCs. As a result, the number of FQHC grants tripled under his administration. The program was further expanded under the Affordable Care Act, in which more than $12 billion was allocated to FQHCs.

The program has existed for decades because of its rare bipartisan support. In addition to appealing to Democratic idealism about health as a human right, FQHCs provide care to many of the rural regions where Republicans cannot leave constituents to die.

As St. John's continued to grow and open new sites in South Los Angeles, a disproportionate number of trans people began seeking care there—at least in part because the population disproportionately lives in poverty, and 91 percent of patients seen at St. John's live below the federal poverty level. Many patients reported having acquired hormones of dubious quality and safety on the street and administered injections to themselves. Some attempted to mimic the physical shaping of sexually dimorphic features by injecting household materials like bathroom caulk or vegetable oil into their breasts, cheeks, and butts. The body tends to attack these substances, walling them off into hard pellets. People have died when such pellets make their way into blood vessels and migrate to block the pulmonary arteries. One patient came back to a St. John's clinic with recurrent infections as oil pellets slid down under her skin and into her calves and feet and formed aggressive red boils that needed to be surgically drained and treated with antibiotics.

Amid these dangerous practices, St. John's director Jim Mangia saw it as a "no-brainer" to begin providing transgender-specific health care.

In the current accepted parlance of trans-rights-advocacy organizations, transitioning genders is the social, legal, and/or medical process of affirming a person's gender identity. This may involve taking hormones (by mouth or by injection into muscle), undergoing various surgeries, changing names and pronouns, and changing identification documents. Largely because the outward process is predicated heavily on physical transformation that can be accomplished only with hormones and surgical operations, both of which must legally be mediated by a physician, the process of transitioning has becoming the province of health care. As a result, the subject of gender transitioning has tasked the health care system with addressing structural social issues.

In 2014, in collaboration with the Oakland-based Transgender Law Center, St. John's expanded, launching its Transgender Health Program and making its holistic approach to health for the trans population explicit. It began in January of that year, with nine patients who were being provided gender-transitioning hormones in addition to general primary care. Mangia hired a transgender care specialist named Cac Cook, who came down from the Bay Area. Mangia told Cook they expected to grow to around seventy-five patients by the end of the year. Cook laughed and said it's going to be bigger than that.

In its first year of existence, based on word of mouth alone, the clinic grew to almost five hundred.

PART TWO

PERCEIVING

THE FEELING PARTS

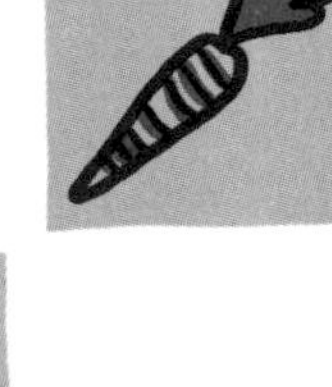
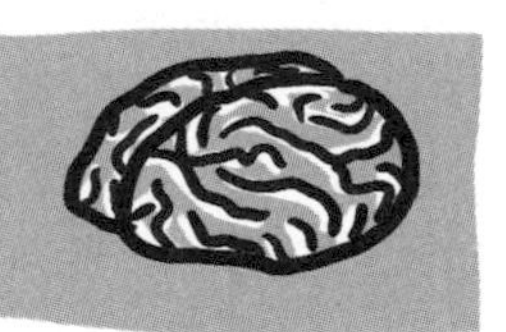

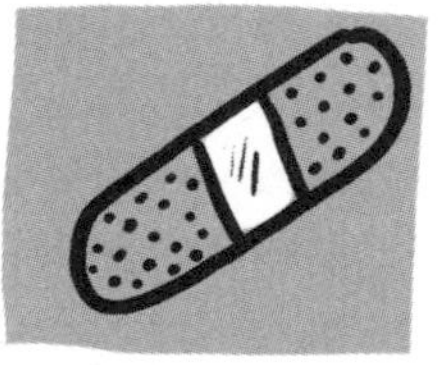

At night, when Kaspar Mossman was in boarding school, he used to tie himself to his bed. It involved a padlock. If he didn't do it, he'd wake up and look in the mirror and think, "Jesus, what did I do to myself?" His face and neck would be red and raw from scratching, still itching in the places where the itch had not turned to pain.

Like most people with chronic itch, Mossman tends to have it worst at night. Especially in his sleep, even though he keeps his fingernails very short, he scratches himself so intensely that his skin tears. He describes his appearance after an episode as if he had taken a cheese grater to his skin.

When he was a teenager, his biggest concern was that his bunkmates were quick to tell him when he "looked like shit." After one late night out drinking, a guy came into his room, presumably to ask about bumming a cigarette, and saw Kaspar tied to the bed. So he did what any teenager would do: He turned the lights on and called everyone in the dorm to come make fun of him. Thereafter, Mossman recalls, "Everybody thought that I was an extremely strange person."

He's more comfortable with that now. "We're all so emotionally vulnerable in our teen years," Mossman, now an introspective forty-two-year-old communications expert with a University of California tech incubator, explained to me. "Well, I shouldn't generalize," he defers. "*I* was emotionally vulnerable. I'd think, 'Where's your place on the totem pole? Are you strong? Attractive? The girls aren't going to dig this stuff. The guys are going to think you just look diseased. You're basically a leper, you're a biblical leper.'"

To make sense of perceptions, in the broadest sense, is traditionally where physiology defers to philosophy. Understanding the world as a collection of perceptions—drawn from the complex sensory systems through which we itch, ache, crave, are attracted, are repulsed, and react to those perceptions based on neural pathways that are set in motion before we're even consciously aware of them—tends to help me keep everything in perspective. I think it encourages empathy. At the very least, it can quell instincts to treat one another like lepers.

What is itch?

Although boarding school and university were the worst years for Kaspar Mossman's lifelong itch, his mother describes him as "red and screaming from the beginning." Even knowing that origin story, his parents grew aggravated at times, taking a disciplinarian approach. Mossman carries a vivid memory of his father sitting him down and looking him in the eye, saying, "You're just going to have to stop scratching. You have to control yourself."

This idea—that itch ought to be managed by nothing more than simple self-control—is a common one, according to Brian Kim, an itch specialist at Washington University in St. Louis. There, in 2011, the Center for the Study of Itch became the first itch-centric study center in the world.

"There's such a bias against assuming that there is a treatable medical cause for itch," says Kim, who trained as a general dermatologist. His job as an itch specialist has become treating the people suffering from often disfiguring-if-not-debilitating itch without a clear medical cause across the country after other doctors have thrown up their arms in defeat. Because it's a medical problem without an easy answer, Kim is sent patients from all over the world. He deals with some enigmatic cases, but also ones that are emblematic of how dismissive people are toward itch.

One woman recently made the trek to St. Louis to see Kim after having suffered from severe itch for a year and a half. She came into his office with excoriations all over her body. She had seen several physicians and, when antihistamine medications had no effect, had been prescribed an array of psychotropic medications. It was a common response: to label itch patients as mentally ill and treat them accordingly. But when she came to Kim, he did a chest X-ray, which is part of a basic evaluation for almost any mysterious condition. The radiologists saw a mass in her chest, which turned out to be lymphoma. She seemed almost relieved when he gave her the diagnosis, Kim recalls. She was somehow vindicated.

"That's not to say that we should assume itch means lymphoma—not at all!" Kim said. "It's just to show that there is *such* a bias against itch. No one wanted to believe that she had a medical problem, especially after seeing so many physicians."

More often the cases that make their way to the Center for the Study of Itch are nebulous, and chest X-rays and most any other conceivable test are done to no effect. Rather, patients present with otherwise clean bills of health and yet still a diffuse, debilitating itch. He calls it *chronic idiopathic pruritis* (*pruritis* = itch, *idiopathic* = of unknown cause). It tends to affect people later in life. By various estimates, chronic itch (defined as occurring persistently for longer than six weeks) affects between 8 percent and 14 percent of the world's population. Instead of dismissing people, Kim does his best to take their itch for what it is.

"It's very hard to define *itch* as a medical condition," says Kim, who believes that half of the benefit that he's able to provide at the center is validation and respect. "The way it has been is that itch is a manifestation of something else that you need to fix. In other words, fix the eczema, don't talk to us about itch." Currently the only FDA-approved medication designed specifically to treat itch is called Apoquel, and it is for dogs only.

(If you're like me, you're thinking, *What's stopping me from taking a medication for dogs?* Often the medications given to dogs and humans are the same—as with the antianxiety benzodiazepines like canine Xanax. Though in the case of Apoquel its label carries the uninviting warning that the drug may "increase the chances of developing serious infections, and may cause existing parasitic skin infections or pre-existing cancers to get worse.")

Kim and his itch team use an array of human-approved medications in "off label" (unproven or potentially unsafe) capacities. First, he tries to shut down the inflammation with steroids. Then he uses aggressive moisturization to heal the skin barrier. Then he adds neuromodulatory drugs and sometimes antidepressants like mirtazapine and amitriptyline. The multipronged approach is critical: "What we know is itch is not simply neuropathic, it's not simply immunologic, and it's not simply in the epithelial barrier, but probably a combination of all these."

Unlike patients that carry labels of idiopathic pruritus, Kaspar Mossman carries a diagnosis of severe *atopic dermatitis,* more commonly known as eczema, the most common cause of chronic itch. Typically diagnosed when a doctor sees red skin that's scabbing and torn from scratching, eczema is actually a spectrum of conditions, and experts are working hard even to characterize it. A unifying feature among people with the condition is that their skin is always at least a little like parchment, even on good days, and those good days are interrupted by flares of red patches that are intensely itchy.

"The worst is when well-meaning people will say, 'Hey, did you know you have this patch there?' Mossman told me. "*Do I know? Get out of my face, I'm trying to hide it*. And they give you a story about how you should be doing yoga or taking fish oil. Even well-meaning people just feed into the problem. When you get a bad flare-up, you just don't want to be around people. I bet that most people with eczema have a reputation as a crabby person. It becomes part of your personality."

An international consortium of experts is still working to agree on a set of metrics to assess eczema and what constitutes a "flare." Occasional studies will report that it seems to be associated with—not necessarily caused by—a certain gene, a difficult task when there is no standard definition for the condition.

So Mossman has learned to manage his own skin over the years, to avoid certain triggers, like liquor and hot spices—he says cayenne is the worst—and hard cheeses. Parmesan contains high levels of the compound histamine, the same that is produced in the skin itself during allergic reaction.

"I discovered this when, at one point, I decided I was going to become a cheese aficionado," Mossman recalls. He began by going to a cheese shop and getting a bunch of cheeses, and then eating them. What followed was one of the most intense of all the nights of itch in his life. "I figured out it was the Parmesan," he says with consternation.

Stress, too, sets off the itch. In school, exams were particularly miserable experiences. Which makes it especially impressive that Mossman went on to earn a PhD in biophysics from Berkeley. It's something he has come to resent. "What a dumb way to learn, going to school. There are tests, and these stupid big papers, and you work and work and cram and stay up late and regurgitate it, and it's this incredible stress level, and you never get asked about it again."

It's a common refrain, though more troubling for a person with a condition in which the physical toll of stress is so immediately, so outwardly manifest. Itch is no superficial novelty, but a complex paradigm of the mind-body interface that factors in everything from everyday quirks to debilitating disease. Itch rather helps us appreciate the tenuousness of any understanding of the self that is predicated on a dualism between mind and body.

German professors once demonstrated that they could get people to itch using purely verbal and visual cues. They did this by placing video cameras around a classroom during the course of a lecture (entitled "Itch: What's Behind It?"). Unbeknownst to the audience, the first half of the lecture was loaded with itch-inducing imagery and terminology—ticks, mites, the word "scratch," and so on. The second half of the lecture was itch-neutral in content. The cameras caught significantly more scratching going on during the first half.

Are you itching now?

. . . Yes, please stop. Also, **why does scratching feel good?**

Scratch-behavior management is part of the approach to treating severe itch. The complex psychology is much more complicated than simply imploring a person, "Don't scratch!," which can be akin to telling a severely depressed person to "turn that frown upside down." Outright dismissal compounds stress by cultivating a sense of blame and failure of will. That makes Mossman itch. Just thinking about itch makes him itch. Thinking about the *absence* of itch makes him itch.

One absentminded scratch can be all it takes to send Mossman into an unstoppable vortex. He tends to continue until he starts bleeding. Then the itch becomes pain. That's also when he becomes susceptible to bacterial infection. His skin gets hot and inflamed, and it'll take several days before it scabs over, heals up, and cycles back again. When itching and scratching keep people awake, the sleep deprivation itself can trigger further outbreaks, in a vicious cycle.

Scratching is one of the few endeavors where we inflict injury upon ourselves and find the experience pleasurable. Even those of us without eczema scratch ourselves hundreds of times a day in a way that is not at all productive. The compelling evolutionary explanation for itch is that it persists for its protective rather than its torturous potential. We itch so that we might identify insects clinging to our skin and remove them before they can impale and inject us with the infectious agents that manifest as malaria, yellow fever, river blindness, typhus, plague, or sleeping sickness. The deadliest animal is still today, by no small margin, the mosquito, which kills hundreds of thousands of people annually by spreading malaria.

For that reason, our sensory systems might do well to reward us with pleasurable sensations from scratching, and to err on the side of caution by sending itch signals at the slightest provocation, however unlikely it is to actually be a mosquito. Mossman describes the hedonic pleasure. "I think I enjoy scratching more than the average person because the itch is so intense," he says. "Probably there's a bit of addiction going on there." This idea is supported by the presence of itch-scratch behavior throughout the animal kingdom, as abundantly evidenced in the viral GIFs and videos that this phenomenon produces, like BuzzFeed's "17 Animals Who Are Experiencing the Perfect Scratch." Which is

to say that some of us so enjoy scratching that we enjoy watching animals scratch.

The properly functioning "itch-scratch cycle" is described in scientific literature, wherein the habit of scratching is distinct from but no less enigmatic than the sensation of itching. The itch-scratch cycle is not a "reflex" by the standard use of the term—where a sensation is transmitted by nerves to a person's spinal cord and, without wasting time going to the brain, is translated into immediate action, like gagging when something is caught in your throat, or extending your leg when a doctor takes a plastic hammer to your knee. People with no brain function at all will extend their knee when the patellar tendon is struck with a hammer; but they will never scratch themselves. People who have had limbs amputated can still experience "phantom itch" in the limb, which suggests that, like pain, itch can originate in the central nervous system. Just as our brains fill the blind spots in our retinas, when we become suddenly devoid of actual sensory input, we essentially guess at what that input would be.

Neuroscience professor at UC San Diego Vilayanur Ramachandran believes these phantom pains and itches can be treated by reorienting the person's sense of spatial awareness, which he has successfully done in experiments that use mirrors to replicate the appearance of a lost limb. Ramachandran's mirror approach essentially attempts to press restart on a person's sense of his or her body.

It was only in 2007 that scientists solidified the notion that pain and itch involve fundamentally distinct pathways in the spinal cord. Before that, the prevailing understanding was that itch was a type of pain. In 1997, scientists had discovered nerves in the skin that perceived itch specifically, but it was still thought that this signal merged with pain pathways. It was from Washington University in St. Louis that Zhou-Feng Chen and colleagues announced in the journal *Nature* that they had discovered a neural receptor that transmits itch, and itch alone. The receptors are known as gastrin-peptide receptors. As Chen describes it in his biography on the Center for the Study of Itch's website, beneath a stoic portrait of the eminent scientist in front of a wall of textbooks, this discovery "opened up an exciting new frontier for deciphering itch circuits and function."

In 2014, Kim and Chen were joined in St. Louis by Hong-Zhen Hu, a neuroscientist who came from Houston because he saw a unique opportunity to "integrate biology at all levels." In that way, itch represents an emblematic approach to contemporary medical research. Hu specializes in the skin cells. Kim focuses on the immune cells. Chen is working on the central nervous system. Their colleague Qin Liu is working on the primary sensory neurons, and Cristina de Guzman Strong is working on applied genomics. This multidisciplinary approach is necessary for a problem that in a recent comprehensive review article was described as "mediated by the interplay between epidermal barrier dysfunction, upregulated immune cascades, and the activation of structures in the central nervous system." Which is to say, it's complicated, and we really don't know, but the answer isn't going to be one single abnormal thing.

The lead author of that study was Gil Yosipovitch, who founded the country's second comprehensive itch center, the Temple University Itch Center, in 2012. The dermatologist and neurobiologist has been called "the Godfather of Itch" by no less than National Public Radio (he does not identify himself as such in conversation). He is deeply aware that his name contains the word "itch," but is politely affirming when people notice and remark about it to him, in my limited experience. Yosipovitch has published dozens of articles in scientific journals on the nature of itch, but he is also, like Ramachandran, open to "nontraditional" therapies as part of a holistic approach to itch treatment. In one desperate case, after a young man was beaten with a baseball bat and suffered brain injury that led to interminable itch, Yosipovitch referred him for "therapeutic touch"—a dubious practice in which healers claim to "manipulate energy fields" by moving their hands over a person in various places in various patterns learned in weekend seminars in hotel conference rooms. He was supportive of trying therapeutic touch not because he actually believed it offered therapy, but for its placebo effect and potentially some resetting of expectations.

Yosipovitch concedes that few people with chronic itch could be treated with mirrors. Indeed, many patients, like Mossman, have a less-than-favorable relationship with mirrors. The significance of Yosipovitch's invoking healing touch is an admission that he is unable to explain what was happening inside the badly beaten brain of his patient, or how to help him. Desperate people want to believe they are doing *something*. The Godfather of Itch often cannot treat many patients as effectively as he'd like with the best in biotechnology at his disposal. He was able to help Mossman, also indirectly, by leading him to the eczema community.

Mossman first heard about Yosipovitch and his itch work in a *New York Times* article. The piece started out describing the plight of people with chronic itch. To Mossman's eye, the people with severe eczema came off "as if they were freaks." He decided that itch needed a voice, so he did what any self-possessed Bay Area science enthusiast would have done in his place in 2010: He started a blog. He filled the pages with the latest in research and treatment. Mossman put a thistle on the top because he wanted to illustrate eczema in the abstract, and "prickly" was the perfect representation for his flares.

Nestled in an arid suburban strip mall adjacent to a driver's ed school and a purveyor of organic mattresses in San Rafael, California, sit the offices of the National Eczema Association. When Mossman found out about the organization, and that it was only a thirty-minute drive from his home, he arranged a visit. At the time the staff was two people. It has since grown to six. The association publishes a magazine, the covers of which are less subtle than thistles. The group's purpose is to cultivate both eczema awareness and funds for eczema research and treatment; their website includes templates for donating vehicles, stocks, and securities, and for leaving something to the National Eczema Association in your will. (Want to exist eternally via bequeathment of your estate to the betterment of skin disease but don't want to deal with the paperwork? Just copy and paste.)

The National Eczema Association is a source for advocacy and information, but, more important for Mossman, connection. Not long ago, people growing up with severe itch didn't know anyone like them. They were less likely to spend time in public, and likely didn't want to talk about it. By contemplating and discussing his itch, Mossman has learned to take calculated risks. Even though alcohol is a trigger for him—as it is for many people with eczema—he loves alcohol. So a night out might go something like this, he tells me: "Say I'm in a place where people are having whiskey. So I'll have a whiskey. Or I'll have two whiskeys. Or I'll have four. I'm not an alcoholic, but I will indulge. I'll feel great, but I know I'll wake up in the middle of the night four hours later with a terrible itch. You learn about triggers, but sometimes you're just like, 'What the hell.' You get tired of denying yourself what everyone else is having. When the itch is there, you hate it, and you're consumed by it. And when it's not, it's like it never existed."

Mossman is hyperaware of coming off as self-pitying when he talks about his condition, because everyone itches, and everyone scratches. But when people say that itch is just something everyone has, he can feel his blood pressure rising. "At this point in my life, I don't really care what I look like. I'm married, I have kids, I'm not out there to attract a mate anymore." But the casual dismissal of his symptoms still bothers him. "The overall feeling is 'Wow, I missed out on having a full life.' Am I bitter? Yeah, a little bit. But then I look around and see people in wheelchairs. Yeah, I have a condition. It sucks. It has probably fed into my crabby personality. I didn't really enjoy my social life. But there are people with acne who didn't either. Most people have something going on. Just with eczema it's showing on the outside."

Ultimately the most unique element of this mystery might be how Mossman managed to tie himself to his bed. That's where his memory, which is so crystal clear when it comes to insults and dermatologic humiliation, gets foggier. It involved an elaborate system of loops and twisting. He can't exactly remember how or why the padlock was involved.

He does remember that a year after he was outed as a bed-tier, a schoolmate confronted him in a group conversation, in front of girls. "Hey, remember that time you tied yourself to the bed because you couldn't stop scratching?" The guy was expecting some kind of reaction, but Kaspar didn't give it to him. Inside he felt desperate to be anywhere else. But he smiled and said, "Yeah, that was hilarious."

Can I "boost" my immune system? What is the immune system even, lol.

Robert Gallo, one of the microbiologists who discovered HIV, told me at a virology conference in Granada that the most threatening infectious disease may very well be the flu. He and many of his colleagues believe the odds are good that in the not-distant future there will be another outbreak of influenza that will rival the 1918 pandemic, which killed around fifty million people. Even now the disease caused by the flu virus is perennially among the leading causes of death in many countries.

So it is important to note that Coca-Cola's sugar drink Vitaminwater does not protect against the flu. Which it would be possible to think if you saw an ad for Vitaminwater that showed a bottle of the hypercolored liquid over the slogan "flu shots are so last year."

The ad also claimed "less snotty noses" and "more immunity." Maybe even more interesting than the baldly dangerous flu claim is the complex idea of *more immunity*. It is the same one made by a rapidly growing number of products, largely of the "dietary supplement" variety, that promise to do something to the immune system. Often the word "boost" is used, though sometimes it's "enhance," "strengthen," or "supercharge." In any case, it sounds good, if you don't know what the immune system is or if you don't know what vitamin C does.

Immune-boosting claims are usually based on vitamin C, or ascorbic acid, a chemical easily added to most anything. It comes in a fine white powder, most of which is synthesized in China. The corn sugar sorbitol is fermented into sorbose, and then genetically modified bacteria turn sorbose into 2-ketogluconic acid. Apply a bit of hydrochloric acid, and you have ascorbic acid.

"Ascorbic" comes from "anti*scorbutic,*" because vitamin C was discovered as the compound that could prevent the horrors of scurvy (from the Latin *scorbutus*). In sailors who first managed transatlantic voyages, the death rate to scurvy in the eighteenth century was often around 50 percent. They would bleed from their gums and eyes as they died. Though scientists had no idea at the time, scurvy is a disease of collagen (like Rafi's epidermolysis bullosa). Because the protein is everywhere in our bodies, we fall apart without it. It is constantly being produced anew, so our bodies require an ongoing supply of the compounds used in its production, one of which is ascorbic acid.

For centuries before the discovery of ascorbic acid, sailors observed that while they usually ate no fruits or vegetables at sea, bringing oranges, lemons, and limes on the voyages and sucking on them periodically eliminated scurvy. The sailors may have been so-called limey bastards, but they did not die of scurvy. Something in the fruit seemed to be antiscorbutic. In 1933, ascorbic acid was identified and proved to be amazingly effective at preventing scorbutus. Just a microscopic dose and scurvy is avoided.

Like many of the compounds known as vitamins, ascorbic acid is a *coenzyme* that assists enzymes in speeding up chemical reactions inside our bodies. Like the other vitamins, its presence is vital, lest we suffer horrific disease. The role of ascorbic acid is to help with the reaction that converts a precursor molecule into collagen; with just a microscopic dose every week, there will be plenty of the coenzyme to facilitate the reaction that produces collagen. Take additional vitamin C and you will not make extra collagen. Your kidneys will excrete the excess, usually without complication.

Because ascorbic acid was among the first compounds humans discovered that could clearly prevent a gruesome disease—a disease in which people literally came unglued, bleeding from every part of their body in excruciating pain—it was easy to extrapolate that this must be some sort of miracle compound that would prevent many diseases. If vitamin C can accomplish this apparent miracle, what else might be possible?

The term "immune system" was coined in 1967 by the Danish researcher Niels Jerne. At the time, there were two competing theories of how immunity worked, based on antibodies or white blood cells. Jerne's concept of an immune *system* united the different pathways by which a host can protect itself from disease—by neutralizing not just disease-causing microbes, but any potentially disease-causing substance.

The immune system is an unprecedented concept in modern medicine. Unlike the cardiovascular, gastrointestinal, or neurological systems—which traditionally refer to discrete sets of structures in particular parts of the body—the immune system describes a function throughout the body. It includes the lymph, which courses through vessels that connect the nodes throughout us, and the spleen, which also filters blood and creates antibodies that lead to long-lasting immunity—an ability of our blood to "remember" certain infections and not fall prey to them again. The immune system is also our bones, which produce the blood that remembers and ingests and ignores compounds accordingly. Blood cells act by causing inflammation and oxidation, and by neutralizing inflammation and the products of oxidation. The immune system is the linings of our mouths, throats, lungs, stomachs, and bowels—everywhere that comes into contact with the outside world, and all the cells secreted on those surfaces that can consume and destroy certain substances while harboring others. The immune system is in the skin, not just as the physical barrier to keep pathogens out, but as an active organ secreting molecules that harbor a population of skin microbes that themselves protect us from disease-causing infection.

Especially since the completion of the first phase of the U.S. government's Human Microbiome Project in 2013, an international initiative that determined that our bodies contain more microbes than human cells, the common wisdom that the immune system's job was to separate "self" from "other" has turned out to be a drastic oversimplification. The composition of our bodies is in constant flux, absorbing compounds through our guts, skin, and air, and constantly changing our microbial complements. The fundamental construct of self and other is becoming untenable. The immune system is not altered or enhanced by our microbes so much as our microbes are part of our immune system—as are the compounds that come onto and into our bodies.

The immune system is essentially our entire body, including our microbes.

When Gallo and colleagues discovered the *human immunodeficiency* virus in 1986, Jerne's "immune system" term quickly came into everyday parlance. As AIDS was explained to the terrified public, weakened immune systems were considered decidedly *bad*—evidenced by the pandemic of suffering and death so plainly before us. And so strengthened immune systems must be good. The stronger the better, naturally.

Certain diseases that compromise the immune system are indeed fatal. But the immune system is not a monotonic concept that can be filed under reductive good or bad. Indeed, it would appear that more human disease is the result of immune system *over*activity. Many of what are known as inflammatory diseases seem to be the result of pathogens that are long gone—Crohn's disease, celiac disease, and eczema are all immune responses.

For its part, vitamin C is a coenzyme involved in the reactions that produce collagen proteins. It does not prevent the flu, or even the common cold. This is a pernicious myth that leads people to waste money on supplements that serve primarily to make a person look paranoid about scurvy.

Before investing in any "immune-boosting" products, consider Harvard neurology researcher Beth Stevens. Her work is illuminating how the immune system is involved in learning. In the brain are cells called microglia, which can move around and engulf other cells. They are part of the immune system in the classical sense, long known to help clear debris and waste from the brain, especially in the wake of injury. But recently we've learned that these cells also destroy the connections between healthy, uninjured cells as a person ages.

When we are born, our neurons have branching connections to many neurons all around them. Beginning in the first years of life, those branches start to disappear, as we train our brains to follow certain pathways. This is often referred to as "learning." At a smaller scale, it's called "synaptic pruning." While we acquire certain skills, we lose the capacity to learn others. This is why it's so easy to learn at a young age, and so difficult later. Our own immune systems seem to be in charge of pruning our synaptic trees.

If the human brain is a finely trimmed hedge, it's worth remembering that hedges can be trimmed too far. One gene, known as *C4,* encodes a protein that marks debris to be destroyed. In 2016, Stevens and researchers Aswin Sekar and Michael Carroll reported in the journal *Nature* that a variant of that *C4* gene, *C4a,* is correlated strongly with schizophrenia. The gene encodes a protein that marks neural synapses for pruning, a part of normal learning, specifically in the areas determining cognition and planning. It is unlikely that one gene determines this complex disease, but their overall hypothesis is compelling: A "boosted" immune system essentially overprunes a person's synapses, in this case in a predictable pattern we know as schizophrenia.

A similar process seems to cause Alzheimer's disease. Also in 2016, Stevens and colleagues at MIT and Stanford published a paradigm-shifting study of the disease in the journal *Science,* showing that microglial cells actually appear to cause dementia in mice by systematically targeting and "eating" healthy synapses in the brain. Stevens showed that animals with Alzheimer's have more of the protein known as *C1q,* and, more important, she was able to block that *C1q* so that it could not mark synapses to be destroyed by microglial cells.

Like so many diseases, this seems to be a normal process gone awry. If our bodies couldn't prune synapses into neat pathways, we would not be able to learn. We would not form personalities with entrenched likes and dislikes and ideologies. But too much pruning is also bad. As Stevens put it in *Science,* "Instead of nicely whittling away [at synapses], microglia are eating when they're not supposed to."

This is an effect of the immune system that is not advertised on Vitaminwater.

It is probably good, then, that there is no compound that will *boost* that system.

Flu shots, though still less effective than ideal, remain one of few ways to prevent a disease that kills thousands every year and stands to kill many millions. The misdirection by advertisers—that "flu shots are so last year"—is not harmless ad copy. And even the subtler claims made by peddlers of vitamins and juices and tonics that claim to boost the immune system are dangerous, not just in that they imply alternatives that do not exist, but in that they perpetuate ignorance of what immunity *is*. Insofar as there is an "immune system," orchestrating it will be the central tack of medical science in coming decades. It holds the potential to cure cancers and treat dementia and undo genetic anomalies. That will not likely be delivered in the form of a beverage.

How do vaccines work?

Most vaccines are dependent on *toxoids*. That's a daunting word, but all it means is a dead version of a bacterial toxin. It is innocuous. Hence it is used in inoculations (vaccines), which are harmless exposures to dangerous things. It's like getting over a fear of birds by watching a bird documentary, instead of by crawling into an ostrich's nest and messing with the chicks. No one loves bird documentaries, but the alternative is worse. Like bacterial vaccines, viral vaccines involve a small amount of a virus (usually in a form that is dead). This allows the immune system to see and learn and remember these dangerous entities so that they do not take us by surprise later.

Does caffeine make me live longer? I read that it does.

"Health is cool, but isn't, like, kicking ass more cool?" The crowd cheered, though I didn't immediately appreciate that the claims were mutually exclusive. "Fortunately, kicking ass is not a drug claim."

At a trade conference in San Diego called Longevity Now in 2014, the entrepreneur Dave Asprey explained the benefits of his "upgraded coffee" while lamenting the limitations imposed upon him by the government. From a stage before an audience of several hundred, in cargo pants and red-tinted glasses, he said that he couldn't go into specifics because it's illegal to make health claims about coffee. In his specific case it almost is.

"Isn't it weird that in the U.S., if you believe in something so much that you *make* it," he says, "that you aren't allowed to say what it *does*?"

Indeed, some argue that the sellers of bioactive substances such as Asprey are precisely the people who should *not* be informing the public as to the health effects of said substances. He and his "coffee" offer us a master class in how to market a product without technically violating a law. He calls his coffee Bulletproof, and he sells it through a company of the same name. The product's slogan is "Bulletproof Coffee: Supercharge Your Body. Upgrade Your Brain. Be Bulletproof." The company's claims are all technically legal because they avoid promises about curing or treating specific diseases. Supplement manufacturers are allowed to make claims about what a product does, without proof that the claims are true, but they cannot say that it *prevents* or *treats heart disease,* for example, only that it "promotes heart health." It is not legal to claim that a supplement product *prevents osteoporosis,* but fine to say that it "maintains strong bones." These "functional" claims of "general well-being" are rife with suggestion, but it is up to the consumer to make that cognitive leap.

The same rules apply to caffeine, since it is not legally regarded as a pharmaceutical, despite being the most consumed stimulant in the world. Coffee is the product of concentrating a psychoactive chemical from the seed of the coffee cherry. The chemical mimics reactions throughout the body that normally occur in intense situations. When we sense danger, the pituitary gland activates the adrenal glands to secrete epinephrine, or adrenaline, into our blood. Adrenaline is the hormone that's meant to be released when we are under stress and need energy, say, to outrun a bear or lift a fallen boulder off our climbing partner. (He's probably not alive anymore, but it's worth checking.) Caffeine can similarly improve athletic performance in the short term, from the height a person can jump to how fast a person can swim.

The hormone surge also creates a buzz. Part of what's needed to lift that boulder is a flood of energy to fuel our muscles, but part of it is also altering our cognition so that we think we can lift the boulder. This is the psychoactive component of caffeine that makes anything seem possible when brainstorming in one of the modern opium dens that we call coffee shops.

Caffeine works, maybe counterintuitively, by blocking communication between neurons in the brain. It inhibits a chemical called adenosine, which transmits signals across the synapses between neurons. Adenosine slows down our neural activity, allowing us to relax, rest, and sleep (those great enemies of progress). Caffeine could be said to cut the brake lines of the body.

Eventually, if we don't allow the body to relax, the buzz turns to anxiety. Many of us stimulate that fight-or-flight response not in occasional dire circumstances, but daily, in our offices, out of habit and performance enhancement and boredom. Eighty-five percent of U.S. adults consume some kind of caffeine most days, with an average daily dose of 200 milligrams (roughly 18 ounces of coffee).

Today, we drink so much caffeine that people have begun selling an antidote. Rutacarpine is a compound that seems to speed the rate at which cells metabolize caffeine. (At least it did so in caffeinated lab rats.) The drawback seems to be that rutacarpine is long-acting and must be taken regularly, as it takes time to work. This might make you break down caffeine more efficiently, but you'd do so at all hours, always. In Dante's book, I think, this was an outer circle of hell.

Asprey's "enhanced coffee" product is mixed with butter ("grass-fed, unsalted") and a triglyceride derived from coconut oil. The latter he sells individually as a product called Brain Octane Oil, which the company claims provides "fast energy," for just $23.50 per bottle. He assures us it's "cleaned, extracted, and bottled without use of harsh chemicals." While he can't advertise that his Brain Octane Oil can prevent memory loss, he can say that it is "the top choice for reaching peak brain performance."

"We can't tell you what we want to tell you, because then we're selling drugs," Asprey said onstage, rolling his eyes. He nonetheless went on to say that his enhanced coffee improves exercise performance and brain functioning and "grows muscle."

When companies sell a substance as a "supplement" instead of a pharmaceutical, as Asprey does, they can go right to market and say that their product can favorably improve or enhance any bodily function you want. They just need to tell the FDA the name of their product and provide a business address. Because if people start filing grievances or dying, one of the twenty-five people at the federal agency that oversees millions of supplement products may need to track the company down.

Like so many who operate in the $33 billion supplement industry, Asprey has become a master of implying without overstepping. "Coffee drinkers live longer than non–coffee drinkers. Could this be an antiaging substance?" he continued, raising his voice before maxing out on sarcasm: "Shocking!" The audience laughed. Within two minutes, he had managed to get them on his side. He wasn't some guy hawking them buttered coffee; he was liberating them from tyranny. He wasn't a power to be vetted; he was a freedom fighter, offering the information that the powers that be didn't want them to know. It was sublime.

There is, though, a way for Asprey and other producers of bioactive substances to make the kind of claims he wants to. They just have to have their claims vetted by the FDA. Their substance would thereafter be regulated not as a supplement but as a drug. The process takes about a decade and costs around a million dollars, and it may end in a denial if the claims don't prove to be sound.

And so we have the supplement industry.

In the early 1990s, after a spate of death and infirmity among people taking supplement products, the federal government attempted to regulate the industry to ensure some standard of quality and purity for the substances it sells, possibly even to require that the products do the things they claim to do. But the industry unleashed a massive lobbying campaign to block the attempt, including television commercials in which Mel Gibson was held up by a SWAT team in his own home, the government come to take his vitamins.

When you want an angry mob to do your bidding, appealing to fear of government overreach is successful approximately 100 percent of the time. In this case not only was it successful, but the industry was able to capitalize on its momentum, creating and lobbying aggressively for the 1994 passage of one of the most consequential laws in health. Called DSHEA (the Dietary Supplement Health and Education Act), it prevents almost all regulation of chemicals sold as supplements. They are not required to be tested for quality, safety, or effectiveness before going to market.

The DSHEA law also expanded the definition of "supplements" into the realm of the arbitrary. It includes far more than the thirteen molecules we know as vitamins, but also enzymes, minerals, amino acids, herbs, "botanicals," "glandulars," and organ tissues. Most any substance can be included here. The distinction "supplement" used to be a concept that derived from its origin: a chemical that came from a food or was analogous to a compound found in food. Thus vitamin C, to supplement the kind you get from fruit. Today almost all supplement products (including vitamins) are synthesized in labs, and the final products resemble no food. Even the products that do contain chemicals found in foods are mixed in infinite arrays of doses and combinations and contexts.

At a gathering of the past six commissioners of the U.S. Food and Drug Administration that I attended in Aspen in 2016, the chairman who was at the helm when DSHEA passed, David Kessler, was asked why the supplement industry is still today unregulated. "We tried," he said, resigned. Kessler's legacy is reining in the tobacco industry in the 1980s, some decades after the product was proven to be the leading preventable cause of death in the country. "Supplements make tobacco look easy."

Meanwhile, within the pharmaceutical industry—whose public image is somewhere between horrible and deplorable—products must show evidence of safety and effectiveness before going to market. The process is far from perfect, but it is a long, costly endeavor that is meant to protect consumers. Pharmaceutical companies are able to make only very specific claims about what their products can do, and they must clearly publish litanies of potential side effects. (This is why the commercials make the products sound like something not to be enjoyed in his-and-hers bathtubs, but rather to be locked in a sealed container and placed on a rocket to the sun.)

When we combine these vitamins into "multivitamin" products, the result is a formula that no longer resembles anything found in foods or the human body. I'll continue to refer to them as "supplements" because that is the common phrasing, but let's acknowledge that it's a meaningless word that refers to nothing so much as a parallel pharmaceutical industry—one that has accomplished the spectacular feat of selling billions of dollars' worth of most anything it likes, in almost any way it chooses, promising people anything conceivable about their bodies, and yet is not reviled, but instead adored and protected. This duality is extremely important. It gets right to the core of our foundational understanding of health and our bodies.

Asprey did make the factual claim that coffee is "the number one source of antioxidants," which is the crux of many people's belief in coffee and its implications for health. And for some people it's true that they get more antioxidants from coffee than anything else they ingest, though that's maybe less a virtue of coffee than a testament to our antioxidant-poor diets. Even still, antioxidant supplements have not shown a correlation with health or longevity.

Figuring out why drinking coffee—or any other allegedly life-extending practice—works, and even if it does work, involves speculation. Is it the antioxidants? And if so, why would antioxidants be effective when they come in coffee, but ineffective when we take antioxidant supplements in pill form? Antioxidants represent a vast spectrum of substances. Vitamin E is an antioxidant, and taking vitamin E supplements has been shown to increase a person's risk of prostate cancer.

Still, we do hear frequently that drinking a small amount of coffee can be good. This always comes back to the evidence that among long-lived, healthy people, drinking some coffee is a common factor. News stories tend to interpret those studies optimistically, reporting that coffee is good for you. It's more fun to tell people what they want to hear. And it's an interesting correlation. But I've never talked to a doctor who recommends that people start drinking coffee as a health measure—only that if you do drink and enjoy it, that's probably fine and maybe even beneficial. Randomized controlled trials are extremely difficult in nutrition, as the effects of dietary changes are complex and often take years, if not lifetimes, to reveal themselves.

It's easier just to say your product kicks ass.

If there is an effect of coffee on longevity, it is something more global than the potential effect of antioxidants. Such as that constant exposure to the stimulant, even at low levels, suppresses appetite (in a world where most people eat more than is ideal). Or that it encourages social interaction—it inclines us to go out and do things. These are legitimately beneficial results.

But as with all chemicals, the effect will depend on how we use them.

Do we still not know if cell phones cause cancer?

"Hi Colleagues," begins a message that circulated widely in a mysterious chain email in 2010. "I don't know how true this is but just take precaution. Please don't attend to any calls from the following numbers: . . . These numbers come in red colors. U may get brain hemorrhage due to high frequency."

The hoax was spotted by many reasonable people who do not trust anyone who uses the abbreviation "u." Upstanding citizens took to forums to warn others and explain that phone numbers do not change the energies coming from a phone. Crisis averted, but other stories of this nature remain more difficult to parse. In May 2015, for example, Berkeley, California, became the first city in the United States to mandate that companies inform cell phone owners that there is a risk the devices may cause cancer. Later that year in *Playboy,* celebrity neurosurgeon Sanjay Gupta said that he uses a wire earpiece when he's on the phone because "it keeps the radiation source away from my brain."

The World Health Organization and the U.S. National Cancer Institute do not acknowledge any relationship between cell phones and cancer. The International Agency for Research on Cancer once said that there was "possibly" a relationship—insofar as radiation does cause mutations in DNA, which can lead to cancers, and cell phones do emit radiation. So does the sun. So do other people. It would be rather irresponsible to say of anything in the universe that it cannot "possibly" cause cancer. So even though the cell phone question is an old one that no credible body has suggested should give us pause, the underlying concept will come up again and again forever. Whenever technology meanders into our lives (especially in places near our brains or groins), there will be a public airing of concern: "Is [new technology] causing cancer/diabetes/autism/etcetera?"

Scurvy, which was unseen until humans boarded maritime vessels for long times at a stretch, could well have been blamed on the technology of ships.

On cue, coinciding with the release of the Apple Watch in 2015, an article in the *New York Times* suggested that the gadget could cause cancer. The original headline was "Could Wearable Computers Be as Harmful as Cigarettes?" (More plausible if you smoke them.) The concern relied heavily upon the words of a multimillionaire entrepreneur named Joseph Mercola, who sells supplements and runs a blog. Formerly a practicing osteopathic doctor (DO), he takes to the Internet to warn his audience against vaccination and most other earthly things. He has been chastised multiple times by the FDA and Federal Trade Commission for making fraudulent claims. In 2016, he was punished for selling tanning beds that he claimed would not cause cancer and could prevent aging. Five years prior, he was rebuked for selling the "newest safe cancer screening tool" (from his aptly named Dr. Mercola's Natural Health Center), which he publicized on his blog as though he were reporting news: "Revolutionary and Safe Diagnostic Tool Detects Hidden Inflammation: Thermography."

Mercola claimed that his cameras could diagnose "immune dysfunction, fibromyalgia, and chronic fatigue," as well as "digestive disorders: irritable bowel syndrome, diverticulitis, and Crohn's disease," and "other conditions: including bursitis, herniated discs, ligament or muscle tear, lupus, nerve problems, whiplash, stroke screening, cancer and many, many others."

It's true that thermal imaging can detect areas of inflammation—a sprained ankle, for example, would be receiving a lot of blood, and it would therefore be emitting more thermal energy. But that is all it can tell us. It is the crudest of tests. The best that can be said of thermography is that it is "safe."

Mercola was also caught selling illegally marketed substances (namely, dietary supplements) in 2006, including his Vibrant Health Research Chlorella XP, Momentum Health Products Vitamin K2™, and Cardio Essentials™ Nattokinase NSK-SD, which included claims like "inhibits cancer cell growth," "prevents heart attacks, strokes, and blood clots," and "lowers blood pressure." Because these claims go beyond the allowances of the DSHEA law—to the point of outright claims of treating disease—the FDA was able to send him a sternly worded letter asking him to please stop selling them (in which the agency noted that it was not conducting a comprehensive review of his operation, and that if he was selling any other illegal compounds it hadn't noted, he should please stop selling those as well).

Mercola's sales techniques resemble those that have proven effective by demagogues throughout history, fostering fear and then selling reprieve. Step one in selling substances that "neutralize or remove poisonous substances from the body" is establishing that such substances exist (and can be removed). This is best done with certainty and authority. (The article on Apple Watches was later updated with extensive qualifications.)

In reality, there are so many variables in our lives that it's almost never possible to make a categorical recommendation about how much of what substance is clearly dangerous for all or even most people. It's easy for someone like Mercola to suggest that something is dangerous, in part, because it is impossible to prove beyond a shadow of a doubt that anything is *not* dangerous. (As a matter of epistemology, science cannot prove a negative; I cannot prove that emojis do not cause cancer, I can say only that there is no evidence so far to suggest they do.)

Our concerns will flow and occasionally ebb, while things that we know to be harmful go unaddressed. The more interesting and immediate way that phones affect our health is behavioral. We walk out in front of cars and die now at unprecedented rates. We text and drive and kill one another. These are much bigger risks than a tumor, and they warrant immediate attention.

As we continue to merge our bodies with technology—not just glasses and prosthetic joints and dental fillings, but our phones and beyond—the potential concerns could begin to occupy us full time. Among those concerns, cancers can feel quotidian. More pressing: As technologies change what it means to be human, are we adopting them with the right amount of care and deliberation? Are we attempting to make sense of how they've changed us?

This is the realm of Jesse Fox, a psychologist who studies how communication technology affects our sense of self. She describes herself as a gregarious southerner who likes to talk to strangers. At Ohio State University, she directs the Virtual Environment, Communication Technology, and Online Research lab.

"Social networking is so fascinating because it goes back to what we call in our nerd speak 'affordances,'" she explained to me. Those are the properties of technologies that make interactions vastly different from the ways we interact face-to-face. For one, there's our need for social validation.

"We know that when people receive compliments, even on trivial things, there's a positivity effect," Fox said. "But social media has changed that game, because it makes validation accessible 24/7. At any point when I'm needing social validation, I can put something on the Internet and get my fix. It's a question of something that is fine in small doses, but when we start having mainline feeds, we get into some problematic territory.

"We're all kind of blind to how immersed we are personally," Fox added. "When I hear people complain about how immersed other people are, and how much time other people spend on their phones, I'm like, 'Have you self-reflected lately?' We have blind spots for our own behavior."

Then she said some other stuff, but I wasn't really paying attention.

Why do ears ring?

When journalist Joyce Cohen ventures out of her Manhattan apartment, she wears commercial-grade noise-blocking earmuffs. She likens her look to that of an airline baggage handler.

"If your book will rank body parts in terms of suckiness," Cohen suggests, "you can put ears at number one."

Body Parts in Order of "Suckiness"

1. Ears*

Cohen has a little-known condition called hyperacusis, where everyday sounds are perceived as unbearably loud. This is sometimes confused with misophonia ("hatred of sound"), or called selective sound sensitivity. As she describes misophonia, certain noises—especially visceral bodily sounds, like chewing or gurgling—trigger not mere annoyance, but "an instantaneous, blood-boiling rage."

Others report that specific sounds trigger "sadness, panic attack, indecision, loss of cognition, physical itching or crawling sensations, urge to flee, or fight." That's according to the online Selective Sounds Sensitivity community of 5,698 people. It is moderated by audiologist Marsha Johnson, who named the disorder in 1997. The symptoms tend to lead to psychiatric diagnoses: phobic disorders, OCD, bipolar disorder, anxiety.

Cohen and many others in the hyperacusis and misophonia communities are convinced that it is inappropriate to conflate these conditions with mental illness. She highlights the work of the University of Texas neuroscientist Aage Møller, who believes that misophonia is a "physiological abnormality"—something intrinsic to the tiny hairlike nerve cells and circular canals of fluid in the ear, or to the tiniest bones in the body, those just behind the eardrum. Sound waves cause that membrane to vibrate,

* Tied for number one: bowels, skin, brains, testicles, joints.

which moves the bones and sends waves through the canals, moving the hairs that translate that motion into an electrical signal sent through nerves into the brain to be "heard." Anything along this delicate course could go awry. It is important to Cohen and the community that it be known that the pathology is along this course and not in the brain.

To some degree, sounds can trigger visceral reactions in many people. In the brains of those without misophonia, the word "moist" (hereafter referred to as simply "the Word") is a universal example of a phenomenon called *word aversion*. The Word is among a small group of widely hated words ("slacks," "luggage") that are—unlike gruesome descriptors or hate-charged epithets—innocuous in meaning. Twenty percent of Americans say they equate the Word with fingernails on a chalkboard. In a research collaboration between Oberlin College and Trinity University (deemed "an initial scientific exploration into the phenomenon"), psychologists specifically tried to determine what drives visceral aversions to the Word. They hypothesized that certain sound combinations may be inherently unnerving to the brain—in the case of the Word, "oy" juxtaposed with "ss" and "tt." They also suggested that we deem words aversive when speaking them engages facial muscles that correspond to expressions of disgust. They found that aversion to the Word was associated with age, neuroticism, and a propensity to find bodily functions disgusting—but not to individual sounds within the Word itself.

Cohen now works with Hyperacusis Research, a nonprofit dedicated to researching noise-induced pain. Distinct from misophonia, hyperacusis is when everyday noise begins to be perceived as loud and causes pain. At an otolaryngology conference in Baltimore, the group shared the stories of several people with hyperacusis, including part of a suicide note from one thirty-six-year-old former musician: "Today I was on the subway with earplugs in my ears. The person across from me was wearing an iPod. I could hear the music from the person's headphones through my earplugs and it was too loud. . . . No one could ever possibly understand this thing that has happened to me and the utter despair, grief, and sadness I feel all the time. There is never a good night, a good day, a good weekend, a good vacation. It is just torment. Every place I go is too loud."

Like traditional hearing loss, essentially the inverse condition, hyperacusis is also often the result of loud noise. Cohen believes it's important to consider noise a "toxin," the most dangerous of which is the loud noise that we deem tolerable (and so don't avoid). In traditional hearing loss, the microscopic inner-ear hair cells are destroyed by the powerful vibrations produced by loud sounds, usually in gradual spurts, over years. That damage can also cause phantom sound, known as tinnitus (often perceived as "ringing in the ears"), which may be worse than the loss itself. Cohen likens attending a loud concert to "subjecting yourself to assault and battery." She believes that this extreme imagery—technically accurate on a microscopic scale—is warranted for an epidemic that is painless until it's too late. Loud music is like staring at the sun, if staring at the sun were something people did socially.

Tinnitus is the leading cause of disability among military veterans, as well as a common cause of suicide. Quantifying the extent of the relationship is difficult, because the tinnitus and its resulting psychological torment—including social isolation and sleep deprivation—are often diagnosed as psychiatric conditions. But some cases are clear. A fifty-eight-year-old Welsh skipper apologized to his family in his final note that he was "literally driven mad" by the ringing in his ears. A London guitarist gave an ultimatum to his psychiatrist the last time they saw each other that he could not continue with the ringing and was "prepared to be deaf or dead." In the Netherlands, professional clarinet player Gaby Olthuis, who heard twenty-four-hour "screeching," publicly pleaded for medical euthanasia, which she received.

The sounds often seem to be the result of the brain's backfired attempts at filling gaps in auditory input. It is a "phantom sound" similar to the illusion that fills blind spots in our fields of vision, or the phantom pain and itch felt in amputated limbs. Based on this understanding, Daniel Polley, a tinnitus patient and audio-perception researcher at Harvard's Massachusetts Eye and Ear Infirmary, believes that there may be a way to reprogram the brain to stop perceiving phantom sound. Similar to Ramachandran's mirrors, which train people's brains to stop producing pain and itch, audio pathways can sometimes be reset. Polley, who blames his own tinnitus on years of imprudent headphone use, is helping people by using music therapy. With an intensive testing process individualized to each person's hearing loss and tinnitus pitch, he removes specific frequencies from music and prescribes it. Relying on plasticity of neurons to form new connections that essentially ignore the ringing frequency creates a sort of intentional, purposeful blind spot.

Audiologist Allen Rohe has seen this sound-training therapy work. One of his suicidal patients, after a year of therapy, came to experience moments of complete silence.

Can I stop wearing my glasses if I eat enough carrots?

During World War II, the British Royal Air Force spread the rumor that their pilots had super night vision because they ate carrots. To many in the rationing, vitamin-obsessed public at the time, this was not unreasonable, though in fact it was an attempt to conceal that the British had adopted a technology that actually allowed them to see through darkness, known as radar.

Like so many farces, it was rooted in abiding truth. Carrots contain beta-carotene, which our bodies convert into a chemical commonly known as vitamin A. It is necessary for vision. In the cells of our retinas known as rods, there is a pigment called rhodopsin. When light traverses the eye and hits the retina, it bleaches that pigment. The intensity of the light determines the degree of bleaching, which determines the intensity of the signal transmitted to the brain. (These rods can be manually stimulated by rubbing our eyes, which causes a sensation of bright spots even while the eyes are closed.) The pigment cannot stay bleached, though. The rhodopsin must be quickly recycled, like shaking an Etch A Sketch, so that it can be bleached again. This process requires vitamin A. Without it, the rods will remain bleached, leading to blindness.

Deficiency of beta-carotene and vitamin A is often referred to as "night blindness" because the symptoms are first noticed in low-light conditions, but eventually it leads to total blindness. During the U.S. Civil War, around eight thousand Union troops developed night blindness due to vitamin A deficiency. It is still today a leading cause of blindness in many countries, especially in children—even though it is entirely and easily preventable.

However, having extra vitamin A will not make the pigment recycle faster. Take all the vitamin A you want, drink all the carrot juice at the club, and it still won't help your vision. The known limit of human visual acuity is 20/8 (where a person can read from twenty feet away what the average person can read from only eight feet away), and the limiting factor among nondiseased eyes is the number of cone cells in a particular spot in the retina, not the presence or absence of a vitamin. Cone density varies enormously from person to person, based largely on genetic predisposition, from 100,000 to 320,000 per square millimeter among healthy people. As the journalist David Epstein details in *The Sports Gene,* high densities are disproportionately common among professional baseball players. Cone density is one of the strongest predictors of baseball success, and it is beyond anyone's control.

While extra beta-carotene may not improve vision, it can have the effect of turning your eyes and skin yellow. The Office of Dietary Supplements of the National Institutes of Health notes that vitamin A overdose "is usually a result of consuming too much preformed vitamin A from supplements." Juices, too, make it easy to get too much beta-carotene. Juicing strips vegetables of fiber, removing the mass that fills your stomach and triggers a sense of fullness. Just a half cup of raw carrots has 184 percent of the daily recommended amount of beta-carotene, a plausibly safe and reasonable dose. After eating that much carrot, most of us think, "Okay, that's plenty of carrots for me."

A half glass of carrot juice, meanwhile, has many times more bioactive chemicals. A Whole Foods brand multivitamin, likewise, has another 300 percent of the RDA. Do that regularly, and vitamin A will build up in your skin, causing a yellow hue. It should be harmless, though long-term intake of high levels of vitamin A has caused fatal liver failure. When infants are given too much, the pressure inside their heads increases until their intracranial contents are visibly bulging out of the "soft spots" in their skulls, where the cranium has yet to fuse. (Called fontanelles, from the diminutive of *fontaine,* these are little brain fountains.)

All of this excess vitamin A could just as well be going to the kids with night blindness, but no.

How much sleep do I actually need?

One 2015 study of more than ten thousand people in Finland found that the optimal sleep duration that correlated with the fewest sick days (absence from work) was 7.63 hours for women and 7.76 hours for men. So either that is the amount of sleep that keeps people well, or that's the amount that makes you worse at lying about being sick. Or people who were sick with some chronic condition subsequently slept too much or too little as a result of that illness. Statistics are tough. Isolated studies are tough. That's why the American Academy of Sleep Medicine and the Sleep Research Society convened a body of scientists to answer this question by doing what's called a Cochrane

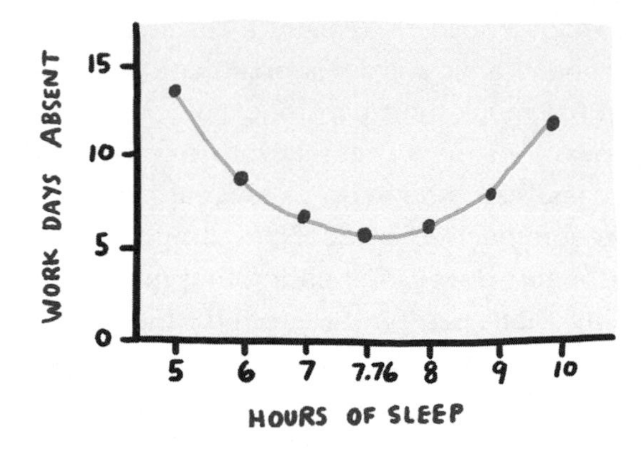

review—a standard approach to trying to reach consensus in science. Sleep scientists from around the world reviewed all known research and looked at the effects of sleep on cardiovascular disease, cancer, obesity, cognitive failure, and errors, ranking each paper on its scientific strength. Then the panel voted on how convinced they were.

Consensus: People should be getting seven hours of sleep per night. When it gets below six hours per twenty-four, there is an increased risk of health problems.

Why do I drool when I nap and not when I sleep?

Salivation is a matter of consciousness. If you're a drooler at naptime, you're a drooler at bedtime. The only difference is that by the time you wake up, the evidence has evaporated.

Should I seriously not be reading my phone in bed? That seems impossible. Why do people give impossible advice?

The United Nations declared 2015 to be the International Year of Light and Light-based Technologies, because light technologies stood to "provide solutions to worldwide challenges in energy, education, agriculture, communications and health." With that in mind, that summer, the New York Blue Light Symposium brought together experts who are trying to reckon with the invasion of this new light habit into our lives. A keynote speaker was Japanese ophthalmologist Kazuo Tsubota, chair of the International Blue Light Society, which he founded in order to "promote public awareness of pertinent research on the physical effects of light" after a 2012 report by the American Medical Association titled "Light Pollution: Adverse Health Effects of Nighttime Lighting."

Of all the things to have health concerns about, nighttime lighting? Well, at least it affords us a good opportunity to reflect on how amazing our endocrine systems are. (Wait, come back!)

When light enters your eye, it hits your retina, which relays signals directly to the core of the brain, the hypothalamus. The size of an almond, the hypothalamus has more importance per volume than any other piece of your body. Yes, that includes the sex organs. Because you would have no sex drive or ability to reproduce without the hypothalamus. Your sex organs wouldn't

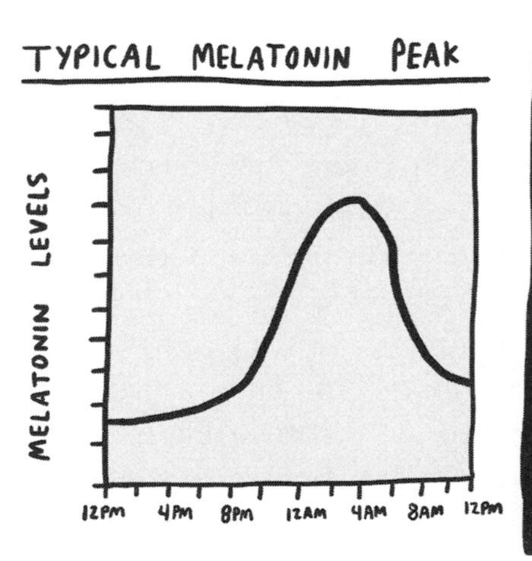

even exist without the direction of the hypothalamus to release testosterone and estrogen in particular ratios. This almond is the interface between the electricity of the nervous system and the hormones of the endocrine system. It takes sensory information from every part of the body and translates it into the body's responses to that information in order to stay alive.

Among other roles in maintaining bodily homeostasis—appetite, thirst, heart rate, and on and on—the hypothalamus controls sleep cycles. The almond doesn't bother consulting with the cerebral cortex, so you are not conscious of any of this. But when your retinas start taking in less light, your hypothalamus will assume it's getting dark outside and you should sleep. So it wakes up its neighbor the pineal gland and says, "Hey, make some melatonin and shoot it into the blood." And the pineal gland says, "Yes, okay," and it makes the hormone melatonin and shoots it into the blood, and you become sleepy. In the morning, the almond starts raising body temperature and blood sugar so you feel less like lying around. That makes us feel colder at night when it gets closer to bedtime, even when the room temperature is unchanged. (Melatonin does this by dilating blood vessels in the skin, which releases body heat.) In the morning, the almond senses light and tells the pineal gland to go back to sleep, which it does. Test your blood for melatonin during the daytime, and there will be almost none.

Melatonin, incidentally, is the only hormone that you can purchase in the United States without a prescription. Under DSHEA, melatonin is considered a dietary supplement and therefore exempt from requirements to demonstrate quality, safety, or efficacy. The pharmacist can't give me the eye drops that help control my glaucoma without a prescription. The pharmacist can't give insulin to a diabetic person without the recurring order of a doctor, to which not all people have easy access. But melatonin, which tinkers with the work of the most critical part of your brain? It's over there in aisle five. Buy as much of it as you like. It's next to the caffeine pills.

Will melatonin put me to sleep? I can't sleep.

One way to fund science is to become one of the wealthiest people on the planet, and then give some of that wealth to advance human knowledge. Bill Gates has taken this approach, funding much study in the realm of deadly infectious diseases and sustainable agriculture. Multibillionaire tech visionary Peter Thiel, who made his first fortune by creating a company that lets people exchange money (PayPal), is only the 638th wealthiest person on the planet. Still, in the spirit of the betterment of humanity, he started a program in 2011 called the Thiel Fellowship. (Named for Peter Thiel.) It is an expansive two-year program, open to "young people who want to build new things instead of sitting in the classroom," according to the application website. "Thiel fellows skip or stop out of college to receive a $100,000 grant and support from the Thiel Foundation's network of founders, investors, and scientists." The competitive opportunity attracts thousands of idealistic young applicants each year, of whom twenty to thirty are selected.

One of those is Ben Yu, a burly-voiced young man who was moved to drop out of Harvard College to venture into the business of selling supplements. In 2015 the Thiel fellow's new biotech start-up, based in Palo Alto, launched a product called Sprayable Sleep. Yu did not find a way to package and sell sleep itself, but he did use his one semester of biochemistry education and the tremendous opportunity afforded him by Thiel's fellowship to create and sell aerosolized melatonin. Spray it onto your largest organ, your skin, and it's supposed to put you to sleep.

When I spoke with Yu, he referred to melatonin not as a hormone but as a "biological signaling molecule." I asked him if that was maybe because customers might be averse to spraying themselves with a hormone. "I thought that might be a loaded word," he agreed, "but it turns out, people don't seem to care." The promise of sleep in a sleep-deprived culture can blind people to questions of prudence. In its initial crowdfunding campaign on Indiegogo, Sprayable Sleep raised $409,798. (That's 2,300 percent of what they set out to raise, from more than four thousand people.)

Sprayable Sleep contains not just the hormone melatonin but also "distilled water from mother Earth." In the FAQ section of Sprayable's site, there is the question "Is it safe?" To which the frequent-question answerer offers, "Very few people exerpience [*sic*] any serious side effects from using topical melatonin sprays."

Unlike melatonin pills, Sprayable Sleep is supposed to keep you asleep, as the hormone gradually percolates through your skin over the course of the night. I tried it for a couple weeks, and I did sleep, but it was tough to distinguish its effect. I sleep most nights. That said, I can confirm that it didn't burn my skin. Also, people don't like it when you pretend you are going to spray it on them.

Melatonin supplements have been shown to make some people fall asleep more quickly, but they aren't proven to increase the total time or quality of sleep. It's like so many things that work in their natural context—in this case, when light tells the hypothalamus to tell the pineal gland to release it—but for some reason don't work as oral (or skin-absorbed) drugs. And of course, as with most things sold as supplements, the effects of long-term use are unknown.

"I'm not sure that anybody's come up with proof that melatonin supplements are helpful," David Dinges, chief of the division of sleep and chronobiology in the department of psychiatry at the University of Pennsylvania, told me. When Dinges was asked to consult to a U.S. National Research Council committee on supplements for the military, he learned that the country spends "a huge amount" of discretionary money on supplements. "But," he said carefully, "no one is quite sure what their value is. It seems to relate to this notion that they can't do you any harm. In most cases that may be true. And maybe, just maybe, they'll help you."

The military's dependence on supplements is no small matter, as journalist Catherine Price noted in her book *Vitamania*. It can be unrealistic to get fresh vegetables to soldiers in the field, so supplemental vitamins are sometimes necessary additions to the hyperprocessed rations that feed the military (MREs). It can be viewed as a potential threat to national security, then, that we purchase most supplements from China. The threat extends to the civilian population, which has become dependent on vitamin-fortified processed foods. The United States does not have an agricultural system that can supply fruits and vegetables to 312 million people, and we do not have the factories that produce vitamin supplements to fortify the processed grains and corn on which we rely. Were the country to go to war with China—or simply stop trading so readily—some Americans who live solely off of processed foods would develop vitamin deficiency diseases like beriberi or scurvy.

What is clear to Dinges is that supplement overuse can be an individual problem too. The fact that the supplement melatonin is a chemical analog of a naturally occurring hormone does little to argue for its safety or effectiveness when taken as a pill or sprayed onto the skin. As Dinges put it, "No child should have a melatonin supplement—or a caffeinated drink—without a doctor being involved. We're talking about adults who might make informed decisions."

The delicate word there is *informed*. Sleep modifiers are left to the whims of Silicon Valley tech companies whose expertise lies in brand engagement optimization strategies. Sleep deprivation, meanwhile, is clearly linked to heart disease and strokes. In Indonesia, a twenty-four-year-old advertising copywriter died after prolonged sleep deprivation in 2013, collapsing a few hours after tweeting "30 hours of work and still going strooong." She went into a coma and died the next morning. A colleague wrote on Facebook, "She died because too much of overtime working, and too much Kratingdaeng attacks her heart." Kratingdaeng is the Thai name of the product known elsewhere as Red Bull.

The vitamin/caffeine/amino acidic concoctions known collectively as energy drinks represent another side of our attempts to manipulate our natural sleep cycles with substances. They have been implicated in a spike of hospital visits in recent years, doubling between 2007 and 2011, according to the U.S. Substance Abuse and Mental Health Services Administration. (The slogan "gives you wings" is perhaps more commercially viable than "attacks your heart.")

"There are many stories about fatalities related to energy drinks, and several lawsuits," Mike Jacobson, head of the Center for Science in the Public Interest, told me. "At least in some people, it has to do with underlying heart defects. When they get this dose of caffeine, they succumb." For now, this is simply a correlation with a plausible mechanism, not yet "proof" of harm.

Caffeine overdose is not known to kill otherwise healthy people directly, but its overuse does alter the body's internal clock, a biological concept known as chronicity. Presumably many of the people who have been hospitalized after consuming energy drinks are also coffee drinkers, notes Jacobson, though few have been made acutely ill by coffee.

The stepfather of the young writer in Indonesia didn't blame her employers, but rather the advertising industry as a whole. Work culture is what drove his daughter to drink the Krating-daeng, he suggested, and to deprive herself of sleep and leisure; a culture where long hours are expected and lauded, where more is better.

Sprayable Sleep is the company's second product, by the way. Its first was Sprayable Energy, which is topical caffeine.

Can I train myself to need less sleep?

As an experiment for his high school science fair in 1964, a sixteen-year-old San Diego boy named Randy Gardner stayed awake for 264 hours. That is eleven days. Since 1964, standards for school science fair safety have changed.

The project was overseen by the Stanford sleep researcher Bill Dement, among others, who took turns watching and assessing the young man's consciousness. By all accounts, the svelte blond Gardner took no stimulant medications. Nor did he seem to suffer many deficits. Dement said that on day ten, Gardner even beat him at pinball.

I asked David Dinges at the University of Pennsylvania how many people could do anything close to that without dying. He confirmed that "when people are sleep deprived constantly they will suffer serious biological consequences. Death is one of those consequences." That said, cases like Gardner's—of people who suffered great sleep deprivation without major setbacks—are also well documented.

There does seem to be a small number of people, sometimes called "short sleepers," who thrive on only four or five hours per night. Dinges said that while the exact number of these people is unknown, "we probably do have people among us—and not necessarily [the commonly cited number of] 1 percent, there may be many more than that—who can actually tolerate sleep loss better than others." This has been shown in studies of people who did transoceanic sailing races, which did not afford them the luxury of long blocks of sleep. The winners tended to be the people who slept the least, often in the form of multiple short bursts.

The concept has spread as people try to apply it to their daily lives. Today a small global community of people practices "polyphasic sleeping," based on the idea that all you need to do is partition your sleep into segments and you can get away with less of it.

Though it is clearly possible to train oneself to sleep in spurts instead of a single nightly block, Dinges says it does not seem possible to train ourselves to need less sleep per twenty-four-hour cycle. Even for the 1 percent (or so) who can survive on less sleep and function well cognitively, he notes that we still don't know how the practice might be affecting metabolism, mood, and myriad other factors. "You may be cheerful, but not cognitively fit. Or you may be cognitively fit, but hard to be around because you're pushy or hyperactive."

Around the time of Gardner's historic science project, the U.S. military got interested in sleep deprivation research: Could soldiers be trained to function in sustained warfare with very little sleep? Their original studies seemed to say yes. But when they put people in a lab to make certain they stayed awake, that wasn't the case. Cumulative deficits accrued with each night of suboptimal sleep. The less sleep they got, the more deficits they suffered the next day. But most interesting was that people couldn't tell they had a deficit.

"They would insist that they were fine," said Dinges, "but weren't performing well at all, and the discrepancy was extreme."

This finding has been replicated many times over the decades, even as many professions continue to encourage and applaud sleep deprivation. In one study published in the journal *Sleep* (my favorite journal name), Penn researchers limited people to six hours of sleep per night and watched the subjects' performance on cognitive tests plummet. The critical finding was that throughout their time in the study, the sixers thought they were functioning fine.

"We don't really track our capability very well," said Dinges, "because we interpret our capability based on motivation, prior knowledge, social entitlement, etcetera."

Effective sleep habits, like everything, seem to come down to self-awareness. During residency, I worked hospital shifts that could last thirty-six hours, without sleep, often without breaks of more than a few minutes. Even writing that now, it sounds to me like I'm bragging or laying claim to some fortitude of character. I can't think of another type of self-injury that might be similarly lauded (except maybe binge drinking). Technically the shifts were thirty hours, the mandatory limit imposed by the American Medical Association, but we stayed longer because, in a hospital, people kept getting sick. You can't just say, "My shift is over. I started yesterday morning, and now it's nighttime, so good luck." No, you stay and help. That proves your work ethic and dedication.

There was always someone new in the emergency room who needed to be admitted, or someone requesting a "sleeping pill" because the lights and noise of the hospital were keeping him awake, or someone on the eighth floor, which was full of the late-stage terminally ill people, who needed me to fill out a death certificate. Sleep deprivation manifested as bouts of anger and despair, mixed in with some euphoria, and some sensations I've not had before or since. I remember once sitting with the family of a patient who had just been admitted to the ICU in critical condition, and we were discussing an advance directive—the terms defining what the patient would want done were his heart

to stop, which seemed likely to happen at any minute. Would he want to have chest compressions, electrical shocks, a breathing tube, or what? In the middle of this, I had to look straight down at the chart in my lap because I was laughing. This was the least funny scenario possible. I was experiencing a physical reaction unrelated to anything I knew to be happening in my cerebral cortex. There is a type of seizure, called a gelastic seizure, in which the person appears to be laughing—but I don't think that was it. I think it was plain old delirium. It was mortifying, though no one noticed.

My experiences were consistent with the findings from the University of Pennsylvania sleep lab: No matter what happened to my body, I never *felt* like it was dangerous for me to keep working. I knew my speech was terse and I was irritable, and I didn't smell the best, but I didn't think anything I did was unsafe. Dinges likens sleep-deprived people to drunk drivers: They don't get behind the wheel thinking they're probably going to kill someone. But as in drunkenness, one of the first things we lose in sleep deprivation is self-awareness. People with the least in reserve show effects the most quickly.

Is it really that bad if I look at the sun once in a while?

Staring at the sun quickly burns the retina, which we don't feel because even though the retina is among the most dense collections of nerve cells in the body, none are the type that perceive pain. Most people know not to do this. What's less known is how many people are blinded by radiation from the sun without even having ever stared at it. Incinerating the retina is not the only way that the sun can damage our eyes.

According to a survey by the Vision Council, Millennials are the least likely generation to report wearing sunglasses "always or often." (Honestly, nothing that I read about Millennials will ever surprise me. Because I am one, and we are emotionless and incapable of surprise.) In their report on these findings, the Vision Council goes on to admonish said Millennials, along with anyone who doesn't properly protect their eyes from the sun. The Vision Council is a registered nonprofit organization, and does things like create consumer-friendly maps of the U.S. cities where ultraviolet (UV) radiation is the most intense (number one is San Juan, two is Honolulu, three is Miami—no surprises. Although you can still get burned in Seattle, remember. Maybe that should be their city motto?)

In 2015 the Vision Council published a glossy "UV Protection Report" titled "Protection for the Naked Eye: Sunglasses as a Health Necessity." Another interesting statistic therein: "While 65 percent of American adults see a pair of shades as a fashion accessory when out on the town, sunglasses are also a critical health necessity." The council also implores us to celebrate "National Sunglasses Day" (June 27, as you may know).

By this point it was unsurprising to see that in the fine print, the group's mission is to "represent the manufacturers and suppliers of the optical industry." So the Vision Council, which sounds sort of like a panel of health experts, is a trade organization. Its authoritative, public-service-toned repositories of health *information* (and not, say, marketing or propaganda) gets top billing in Google searches about UV radiation and eye protection. Googling health information is roughly as reliable for finding objective answers as picking up a pamphlet from the subway floor. ("Why are you doing that cleanse thing?" "I read about it in a pamphlet I found on the subway floor.")

Of course, while this sunglass-selling-first approach makes the Vision Council fundamentally different from a group whose mission is firstly to dispense truth, it doesn't necessarily mean that their information is incorrect. For instance, it is true that when sunburns occur on the surface of the eye, they are called *photokeratitis* (light-induced inflammation of the keratin that makes up the cornea). UV radiation can also cause discolored plaques, called pterygia, to grow on the surface of a person's eye.

Most significantly, though, UV rays cause cataracts. Cataracts are the leading cause of blindness in the world. An even better source of health information, the World Health Organization, estimates that every year, cataracts blind 12 to 15 million people. Only 20 percent of these cases "may be caused or enhanced" by sunlight, according to the WHO. But with the depletion of the ozone layer, more ultraviolet radiation makes its way to our skin and eyes every year. By the estimate of the Vision Council, a 10 percent decrease in the ozone layer can lead to an increase of 1.75 million additional people developing cataracts every year. So, even if that's off by a million, happy Sunglasses Day.

Am I having a seizure?

Beth Usher was a precocious kindergartner, already reading and writing, a ballerina and soccer player. She was the second child of Brian and Kathy Usher of Storrs, Connecticut. On September 23, 1983, just three weeks after the first day of school, the principal called Kathy in a panic and told her something was "not quite right" with Beth. "She's just not acting herself." Kathy raced out of work and into Dorothy Goodwin Elementary.

"I had never seen a seizure before, but I could tell it was a seizure," Kathy recalls. "She recognized me, and she lifted her left arm, but her right arm was just hanging limp. She said, 'Mom,' and she tried to talk, but couldn't."

Beth's episode had a beginning, middle, and end—a defining feature of seizures. By the time an ambulance delivered her to Hartford Hospital, the seizure had passed, and afterward, as most people with seizures do, she felt fine. Doctors had no explanation. Seizures just happen sometimes. This was probably an isolated neurological storm, unsettling but inconsequential, they reassured the Ushers.

To be safe, they wrote her a prescription for the antiepileptic drug phenobarbital. The Ushers plunked her back into her kindergarten class. The phenobarbital made Beth hyperactive, but she was seizure free—for two weeks. Just when Beth and her parents were getting over the shock of the first episode, a grand mal seizure took over the right side of her body. By the time the thrashing subsided, Beth was in an ambulance being rushed back to the hospital in the middle of the night. That's when life changed for the Ushers. Doctors performed a CT scan of the young girl's brain and saw a large area of dead, atrophied brain in her left hemisphere.

Sitting in Union Square in Manhattan in 2015, Brian Usher pulled the films out of a large manila envelope that bore his daughter's name and handed them to me. The Ushers have half a basement full of Beth's medical documentation, as well as news clippings and letters from well-wishing strangers. Brian has a gray crew cut and the bearing of a collegiate football coach. That's fitting, seeing as how he was a football coach at the University of Connecticut, where Kathy worked in research development. Both are now retired. I held the negatives up to the sun and saw a CT scan that showed dark areas of atrophy, wasting of the brain tissue, throughout the left half of little Beth's brain.

This was not just a slightly abnormal CT scan. It was the sort of brain that makes you marvel at the capacity of a brain to be so disrupted and remain capable of operating a human body at all. It is not at all true that we use only 10 percent of our brains, but it *is* true that a person can lead a typical life with little more than 50 percent of the brain intact. Beth is proof.

Pictures of the inside of our bodies are pieces of a diagnostic puzzle. When a doctor looks at a CT scan (or X-ray or MRI), it's most often impossible to say with complete certainty what disease process is plaguing the person. Different conditions can appear much the same, if not identical. A knife that is sitting on a person's chest looks the same in a frontal chest X-ray as a knife

that is lodged in the person's heart. In interpreting a CT scan, a doctor makes a hypereducated guess at the diagnosis. Degrees of confidence in that guess vary from 99.9999 percent to much lower. That educated guess is usually followed by a number of other less likely diagnoses that could present similar appearances. In the case of infections and cancers, for example, making a diagnosis with *absolute* certainty is often impossible without getting a physical sample (a biopsy) of the tissue and looking at it under a microscope. A patient's story is required in order to tell which is the correct diagnosis. We can't go cutting open everyone's heads, so the system works well enough in most cases. Beth's was not one of those cases.

Children rarely have strokes in the traditional sense, but they do endure injuries to their brains during complicated birth. When doctors in Hartford told the Ushers about the dark area in their daughter's brain, they attributed it to the latter cause: Beth's looked like the brain of a child with cerebral palsy—where an area of the brain had atrophied due to a lack of blood flow during labor. What was really going on, however, was something none of the doctors at this hospital, and few in the world, had ever actually seen happen.

Because while the injury made sense in terms of her brain's appearance in the CT scan, it didn't make sense within the context of her story. Children do not simply develop cerebral palsy at age five. Beth's birth had been completely routine. The sort of brain injury that would be so clearly manifest on a CT scan is the sort that could never be caused by an incident that went unnoticed during birth and childhood.

But when the family consulted a neurologist, he agreed that cerebral palsy was the most likely explanation for the darkness in Beth's brain. He said it was probably the cause of Beth's epilepsy. (Because she had had more than one seizure, she now had *epilepsy*.) Cerebral palsy is extremely common, and it looks just like this in a CT scan, so it made sense in a snapshot diagnostic sense. No one treated either issue with the sort of urgency that the Ushers expected, perhaps because epilepsy is incurable, and once gray matter is atrophied, it cannot be unatrophied the way a muscle can. The doctors upped her dosage of phenobarbital, which left Beth running around the classroom "like a lunatic," as Kathy recalls it.

Things became most distressing when Beth, who had always been right-handed, stopped eating with her right hand. She started learning to write with her left hand. Her right foot would not move when she wanted it to. Her seizures became more frequent and more intense. "She kept getting worse and worse," Kathy recalls. "She kept falling. I would take her shopping, and she looked like a battered child. She was black and blue. People would look at me like, 'What are you doing to that child?' "

The panicked Ushers consulted multiple neurologists, only to be assured that Beth had epilepsy, for which there is no cure, simply treatments aimed at minimizing seizures. They altered her medication regimen. They tried Dilantin, valproic acid, and various drug combinations. Her life became trips back and forth to the hospital, where her medication regimen would be tweaked and titrated, before she returned again with another seizure.

By Christmas, Beth was having such frequent seizures that her father had to pull the car over multiple times during the ride to her grandparents' house. Soon Beth was enduring around a hundred episodes every day. Some of them were brief moments of absence that could be dismissed as spacing out. Others dropped her to the floor. In school, her seizures scared the other kids. She regularly hit her five-year-old head. She required constant surveillance.

For Beth and her family, though, there was one seemingly magical reprieve: propping her on pillows in front of the television and turning on *Mister Rogers' Neighborhood.* For the duration of the half hour of public programming, Beth seemed to have almost no seizures. Kathy recalls her daughter talking back to the television: "Yes, I *will* be your neighbor!"

But Beth could not live her entire life in front of a television tuned to twenty-four-hour Fred Rogers. The Ushers were determined to solve the mystery of their daughter's disorder. At that point, four months after the first seizure, they insisted on having a second CT. Instead of waiting for the neurologist to call her in a few days with the results, Kathy marched back and talked directly to the radiologist, who confirmed her fears. The black void in Beth's brain was spreading. Spreading does not happen with cerebral palsy, a static injury. And the seizures grew only more frequent and more intense. Still, no one could explain it.

Then the Ushers took Beth to the University of Connecticut's children's hospital, where pediatric neurologist Edwin Zalneraitis suggested, after a lengthy evaluation, that she might have a one-in-a-million disease called Rasmussen's encephalitis. No one understands its cause, and there's no cure. It means the seizures will never go away, and that half of her brain will simply be destroyed. (A curious thing about the disease is that it leaves the other half completely unharmed.)

Convinced that there must be something that could be done, Kathy called the Epilepsy Foundation outside Washington, D.C. No one there had heard of Rasmussen's. So she made pre-Internet quests to medical libraries across the Northeast. At Yale, she donned a physician's white coat to sneak into the medical library and pull journal articles. No leads. By this point, Beth was having so many seizures that she couldn't go to school. Just when the Ushers thought they had exhausted their options, Kathy got a letter from the woman she had spoken to at the Epilepsy Foundation. It was a clipping from *The Baltimore Sun* about a little girl from Denver named Maranda Francisco. Maranda had Rasmussen's encephalitis and had recently undergone what was touted as a miracle surgery at Johns Hopkins. She was now free of seizures. The surgeon was future presidential candidate Benjamin Carson.

Kathy called Johns Hopkins, where the chief of pediatric epilepsy, John Freeman, asked to see her CT scans immediately. She sent them overnight. Rasmussen's encephalitis can't be diagnosed based on images alone, because it looks too similar to other conditions, so he asked them to come to Baltimore right away. There the Ushers met for the first time with Freeman and Carson. The duo examined Beth for about a half hour before confirming that they believed the diagnosis was Rasmussen's. The diagnosis alone was a news story. A prominent color photo of seven-year-old Beth in a bonnet and colonial-style dress, like an epileptic American Girl doll, adorned the front page of *The Baltimore Sun* in 1987, under the headline "Young Connecticut Girl Learns Her Brain Is Slowly Dying."

But Carson and the team at Johns Hopkins Hospital did not believe that Beth was going to die. Carson was already as famous as a neurosurgeon could be at the time (until that record was broken years later, by himself), having successfully separated conjoined twins in a marathon procedure that Hopkins publicized heavily. He became the youngest-ever chair of pediatric neurosurgery at Johns Hopkins, reputedly willing to perform the riskiest of procedures. As he had done for Maranda Francisco, Carson recommended hemispherectomy: complete resection (or removal) of half of the brain. And he recommended doing it as quickly as possible.

Rasmussen's encephalitis is a disease largely forgotten and

ignored, on which no progress has been made in decades. A child—almost always a child—will suddenly develop severe seizures, lose some ability to speak, and become paralyzed on one side of the body. The disease behaves like a stroke that comes on over months instead of minutes. Eventually half of the child's brain is destroyed—either the right or the left. No one knows what causes the disease, and there is no treatment.

Because Rasmussen's affects only .000017 percent of children, it's what's considered an "orphan disease." These (sometimes simply "rare diseases") are an increasing pool of conditions overlooked by an industry in search of blockbuster drugs like Lipitor or Viagra, something that will be taken by hundreds of millions of people for decades of their lives, something that will be guaranteed to earn enough to justify the millions of dollars spent in research and development.

But a handful of passionate people are trying to shine a light on Rasmussen's. At UCLA, neurosurgeon Gary Mathern curates a "brain bank" with tissue samples from Rasmussen children's brains. His collection currently has thirty-five samples from all over the world, some of which are better preserved and more useful for research than others.

"There are sometimes issues getting human tissue through customs in a timely fashion," relates collaborator Seth Wohlberg, a financial trader who founded and runs a nonprofit called the Rasmussen's Encephalitis Children's Project. After his own daughter, Grace, began intractably seizing and was diagnosed with Rasmussen's encephalitis in 2010, Wohlberg made it his mission to find a cure. Or at least to create some semblance of progress toward an alternative to taking out half the brains of patients.

The fundamental problem, as he saw it, was the lack of unity among the disparate researchers whose work has bearing on the disease. As in so much biomedical research, the interested parties tend to work in silos. The competition that powers the capitalist system of Wohlberg's day job can actually be counterproductive to curing a rare disease. Yet, as Wohlberg put it, bringing scientists together to collaborate tends to go against their nature, "like herding cats."

Even though Mathern now has a catalog of brain tissue to study, he and colleagues admit they have no idea what precisely is going on. The long-prevailing theory is that the disease is a combination of a slow-growing virus and a self-destructive immune response to that virus. (This vague theory is lately being applied to many diseases, of the brain or elsewhere.) Rasmussen's does clearly involve inflammation caused by an antigen, but after culturing, electron microscopy, and DNA sequencing, no one has been able to identify a virus or other infectious agent that could be that antigen.

"We've looked for everything," Mathern said. "If it's a reaction to a bug, it's a bug that's not known to science, and not detectable by the best methods we have."

Nearing retirement, Mathern hopes to leave his mark by figuring out Rasmussen's. But as of now, he is blunt: "We don't know what causes it, but this is a disease that's going to eat your hemisphere." Why does it destroy one half of the cerebrum in its entirety and stop there? "It makes no sense," Mathern mused.

At least now there is a small community of parents and scientists around the world who are able to share their experiences via the Internet, though there is little more to share than existed in 1987. In the face of such uncertainty, no choice was clear for Kathy and Brian.

"When the Ushers went through their situation, they were pioneers," said Wohlberg. "There were no resources, there was no communication. I admire them."

As a hedge fund managing director by day, Wohlberg has been able to contribute more than a million dollars to the cause. It's only through him that Mathern is able to fund the tissue bank at UCLA. "The National Institutes of Health would not invest federal funds in such a rare disorder," Mathern said, "and the free market can't solve this, because no one could afford whatever solution it came up with."

This is the case with so many rare diseases that have no hedge fund managers in their corner. At the time, the Ushers proceeded alone, led by doctors who could not explain why their child's brain was wasting away, offering only the same barbarous surgery that was available fifty years ago.

"You just can't conceive of handing your child over to someone who is going to remove half their brain," Kathy said.

But Beth was getting worse. When Kathy showed her an umbrella, Beth could say, "It keeps your head dry," but she couldn't come up with the word. "She became catatonic at a birthday party," Brian recalls. "She was just completely in a trance."

In her desperate search for a sign that she should hand her daughter over to have half of her brain removed, Kathy even contacted Theodore Rasmussen, the Canadian neurosurgeon for whom the disease is named. While it was Carson who became famous for performing hemispherectomies, Rasmussen pioneered the surgery. The initial outcomes were bad. But later adopters learned that because children's brains were significantly *plastic*—amenable to "rewiring" themselves after surgery or other injury—not only could a child survive having half of his or her brain removed, but the remaining half of the brain could assume some of the work of the missing half. If you had a hemispherectomy at age thirty, after your brain had functionally hardened, Mathern explained to me, "you'd never recover functionally to be worth a darn."

Hemispherectomy for severe epilepsy has been found to halt intellectual decline and even reverse it, with IQs being higher after the surgery than before. And roughly 75 percent of patients are seizure free afterward. But not all hemispherectomies are equal in quality. Insurance companies prefer to force patients to be seen by the nearest neurosurgeon with a scalpel, Wohlberg explains, rather than be sent to places like Hopkins or the Cleveland Clinic, where the most experienced institutions and surgeons can sub-sub-subspecialize to perfect the art. To do otherwise, Wohlberg says, is "a recipe for disaster." Cut too far, just one centimeter beyond the boundaries of the hemisphere, and damage to the brain stem can kill the patient. Likewise, if even a millimeter sliver of brain from the epileptic half is left behind, the seizures can persist just as intensely as if the entire brain were present.

For many people, like Grace Wohlberg, the surgery does not go well. Grace had half of her brain taken out at Johns Hopkins, only to descend into nine months of hell, as her father put it, before Mathern did a second surgery to clear out some remaining brain tissue.

For their part, the Ushers went back and forth from Hopkins three times before deciding to proceed. The turning point was when Beth had a grand mal seizure onstage during a holiday concert at school during "O Tannenbaum." The principal had to carry her offstage in front of the whole school. She was ready for the surgery.

Make-A-Wish Foundation got her a trip to the White House, but because Ronald Reagan was tied up with Iran-Contra, Beth got to meet only Nancy. The industrious Kathy thought of another way to raise her daughter's spirits going into the surgery. She called the studio in Pittsburgh where Fred Rogers taped his show. She explained Beth's unique connection with the show, and that she was "going to have this horrific brain surgery," hoping that someone from the show might send an autographed photo or note from Rogers.

The next day, the telephone rang. Kathy told Beth that a friend wanted to talk with her. This was a rare occurrence, since the seizures had long made friendships difficult for Beth. She took the receiver and said hello, and then so did Fred Rogers. She told him that she wanted to stop having seizures so that the kids in her class would like her. Beth talked with her favorite characters from Mister Rogers' neighborhood (King Friday XIII, Lady Elaine Fairchilde, and Daniel Striped Tiger), becoming momentarily invincible.

The next morning, the Ushers packed up and drove to Johns Hopkins Children's Hospital, where Beth underwent a litany of tests to assure that she could survive the twelve-hour surgery.

It was Carson who greeted the Ushers in the recovery room and told them everything had gone well. He later wrote in his first book that he had been wrong about this. Amid 250 pages of self-deification, the moment of candor sparkles. That night, Beth's brain stem swelled, and she slid into a coma.

In the intensive care unit, her parents, brother, and grandparents kept vigil and tried in vain to will her to consciousness. With doctors bustling and machines beeping at all hours, the Ushers kept a cassette player pumping out Mister Rogers' greatest hits, including Beth's favorite, "I Like You As You Are."

A nurse came and alerted Kathy that she had a phone call, relaying that a man claiming to be "Mr. Rogers" was asking for her. So Kathy went to the nurses' station, and indeed it was Fred Rogers. Over the next two weeks, Fred Rogers called every day to check in.

One morning he asked if it would be okay with her if he visited. Even though Beth was in a coma, unconscious, Rogers flew from Pittsburgh to Baltimore to see her. He carried only a clarinet case. When he entered Beth's room, he opened the case and took out Beth's favorite puppet characters, King Friday XIII, Lady Elaine Fairchilde, and Daniel Striped Tiger. Rogers sat and sang to her. The Ushers have a washed-out three-by-five photo of him leaning over the comatose child's ICU bed, his hands inside puppets.

It would be a great end to the story if Beth awoke from the coma while Fred Rogers was present. Instead he finished his song, rose, and returned to the airport. After two months, a

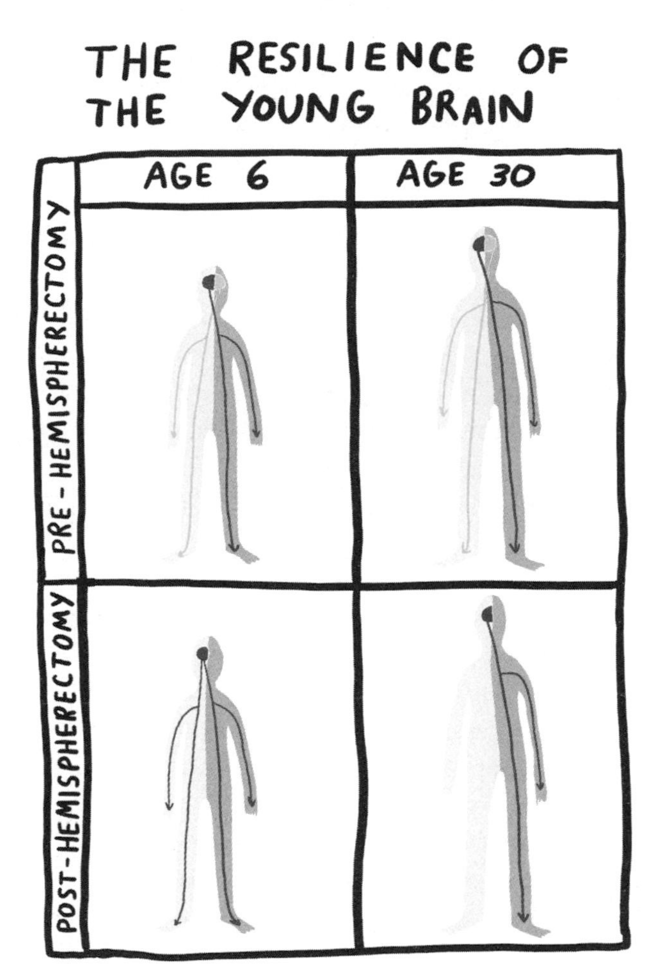

neurophthalmologist examined her and detected so little brain activity that he estimated Beth would never function beyond the level of a newborn.

And then one night, while Brian was lying in a cot beside her bed, he heard a faint "Dad. My nose itches." He bolted upright. He asked if she knew her name, and she said, "Beth Usher."

Do you know where you live?

"Storrs, Connecticut."

Do you know my name?

By that point, Brian recalls, she seemed annoyed. "Brian Usher."

How is laughter medicine?

On a park bench in Manhattan, Beth, now thirty-seven, showed me the yellowing newspaper in which she had *learned her brain was slowly dying*. She explained that her aunt Mary made her a bunch of white bonnets before the surgery, in anticipation of covering the scars to come, and that's why she looked out of place in this century. But today her hair is long, and if anything could be said to be atypical about her it might be the completeness of her presence.

Because the left sides of our brains generally control the right sides of our bodies, Beth should have been paralyzed throughout her right side. But because she was young and her neurons were plastic—her synapses yet barely pruned by her immune system—the left side of her brain was able to learn to control the right side of her body. After just nine months of intensive therapy, she was able to walk again. She does so with a pronounced limp, and her right hand is useful for support, though it lacks fine motor coordination. She has no peripheral vision on the right, which prevents her from driving. She wears a leg brace that she says is supposed to help her walk, but she's not sure that it does. Still, to walk and talk with her is not what one might expect from a person missing a cerebral hemisphere. Her conversational skills are better than those of most full-brained people. "I'm always in my right mind," she assured me, scrutinizing my face for the desired reaction.

Just as she taught herself to walk, Beth believes, she taught herself to be happy. "It was the one thing I could control myself," she said. "I could make my life miserable, or I could laugh."

Her mother says Beth was upset about how she and Brian were always concerned and crying, so Beth would try to make them laugh. She's truly audacious when people are negative around her. She says—not in a bubblegum positive-psychology way, but in an incredulous way—"You can be happier. You're alive."

When people ask her why she limps, she alternates between "I was injured in Vietnam," "I went bungee-jumping off the Empire State Building without a cord," and "My biology teacher did a really strange experiment on me that went terribly wrong." Beth studied humor at one of New York's premier clown schools, New York Goofs.

"These are professional clowns," Brian insists, "so it's pretty intense." The Goof School emerged to fill the training void after Ringling Bros. and Barnum & Bailey Clown College shut its doors in 1998. It now trains many of the Ringling Bros. clowns.

It was there that Beth received her first pie-in-the-face. It was a deeply disappointing experience. The "pies" that clowns throw are tins full of shaving cream, she explained to me, because it adheres to human skin better than whipped cream. These are the secrets you learn only in clown school.

Beth is an avid traveler and collector of sea glass. On one trip to Florida, the Ushers visited the Epcot Center (a midcentury Disney theme park predicated on the idea that the 1990s were going to be really great). In one display about the frontiers of medical science, there was a CT scan, and Beth noticed that in the bottom corner it said her full name: Elizabeth C. Usher. She was shocked. "We had signed a waiver for Dr. Carson," Kathy recalls, resigned.

Mortifying as that might have been to most high schoolers, Beth has always been fine with people knowing that she is missing so much of her brain. She sees her story as an opportunity to turn fear of the unknown, fear of maladies, and fear of physical differences into connection. "The more people who know my story, the less people are afraid of me or anybody with epilepsy," she said. She encourages people not to be afraid of those with seizures, as people often were of her. "You shouldn't just walk away. If anything, you should go hug them."

The subject of laughter as medicine hits home with me. My first academic publication was on the health benefits of laughter. It was a paper called "Humor," and because my coauthor was the eminent radiologist Richard Gunderman, I managed to get it published in the journal *Radiology*. When I'd tell senior physicians about the paper, they would think I was joking. But I was dead serious.

Even among medical specialties, radiology attracts an especially staid type of person—analytically minded and introverted enough to sit in dark rooms all day analyzing images of disembodied illness. Rarely is a human face put to the chest X-ray or CT scan. Our *Radiology* journal article argued that even in a radiology reading room, humor had a role.

For doctors who are disillusioned or unfulfilled in their work but who choose *not* to comb the depths of uncertainty that come with changing careers, there are less drastic options. Anyone, doctor or otherwise, can now become a Certified Humor Professional. Like so many people who devote their lives to humor, Mary Kay Morrison did it as a reaction to a world that didn't seem to make sense. The constantly smiling teacher stayed in the northern Illinois school systems from 1969 to 2005, until she finally got fed up. Morrison recalls being "frustrated" that administrators pushed play out of the kindergarten curriculum in favor of sit-down testing. Her frustration reads as midwestern for what most humans would call *rage*. She heard her colleagues say they had to close the door to their classrooms when they were having fun because the principal might walk by and think they weren't working. "That's the exact opposite of how children should be learning," she tells me. So she started doing workshops on humor for teachers. And then for everyone. She advocates for the positive energy of humor, or, as she calls it, humergy, to promote balance and reduce stress. Her email sign-off assures correspondents that she is "Sending Humergy."

Now the sitting president of the Association for Applied and Therapeutic Humor (AATH), Morrison has designed and implemented a three-year certificate program, as well as a humor curriculum that confers college credit and counts toward continuing education credits for mental health counselors. Graduates earn the distinction of Certified Humor Professional. That, at least, is what goes on their résumés, websites, and LinkedIn profiles. Morrison will refer to you instead as a "HAG," an acronym for Humor Academy Graduate (and, I believe, a pun that playfully subverts the patriarchy). There are currently only twenty-five HAGs in the world. The number, though, is far exceeded by their passion for employing therapeutic laughter.

"A lot of times people think of clowns when they think of fun," Morrison informed me. "We have many therapeutic clowns in our organization, but that's not really the bulk of our mission." The legendary clown-doctor Patch Adams was a member of AATH in his later years, and his work was pivotal to the acceptance of humor in medicine. But the more common usage of therapeutic humor requires no makeup, giant shoes, foam red nose, or abiding sadness.

AATH is open to anyone interested in learning about "the application and benefits of therapeutic humor," for anything from terminal diseases to day-to-day stress. The Certified Humor Professional track draws students from all over the world; they do most of their studies and practice remotely. The annual AATH conference draws more than two hundred people. Many of them came to humor later in life, after the drudgery of professional life wore their spirits to stubs. Harold, a second-year student, is a politician from Norway. Another student is an Australian who flies with doctors to the outback. A Japanese student took up the program as part of a mission to bring humor to Japan, which she considers humorless. In 2015, after her training, she won a Toastmasters competition.

Laughter is proven to release endorphins, like running or consuming opium, and laughter decreases the stress hormones cortisol and epinephrine, thereby improving the function of the immune system. It works even if you just feign laughter, even if you don't find anything funny. The very act of laughter, even devoid of humor, seems to have positive effects on blood pressure and mood. Studies of humor and laughter are small and sparse, which is largely because laughter and humor are not easily or highly monetizable medical interventions. Unlike products that, say, inject bile into one's chin or cauterize a person's heart, pharmaceutical and device-manufacturing companies are not pouring R&D funds into laughter. (It bears noting that we now have a million-dollar robot that can perform hysterectomies.)

Because the mission of the AATH is "to serve as the community of professionals who study, practice, and promote healthy humor and laughter," I asked Morrison what constitutes *healthy* humor and laughter. The definition rests, to Morrison, within the distinction between positivity and negativity. Laughing at someone out of spite might not be as therapeutic as good-natured, cathartic laughter. As I understand it, if you are cackling at human suffering, that is not healthy. If you're laughing because you just noticed how beautiful the world is, or because human behavior is completely incomprehensible and *lol nothing matters,* that's therapeutic.

In 2015, Morrison invited Beth Usher to be the keynote speaker at the national conference of the AATH. Three decades after having half of her brain removed, Beth took the stage singing "If I Only Had a Brain" from *The Wizard of Oz*.

"The place went crazy," her father, Brian, recalled, as Beth shared her message with attendees that, with a cultivated sense of humor, purposefully practiced, a person can endure anything. Her brother was on hand to turn pages for her. She quoted Carl

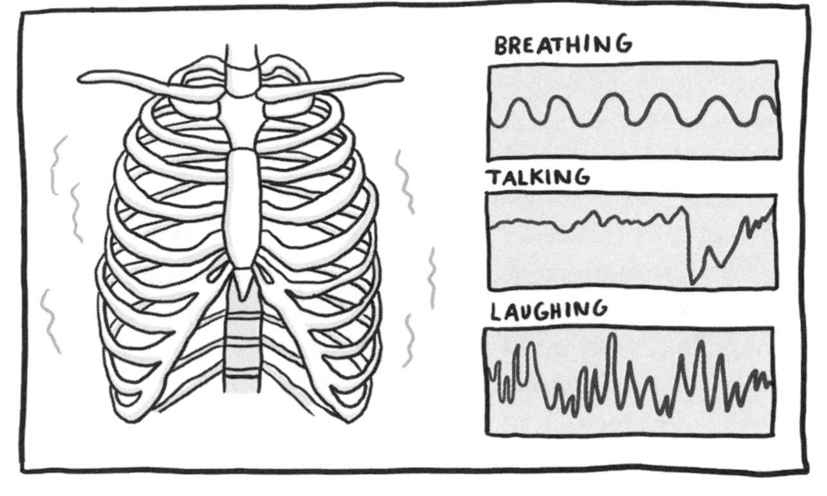

Jung: "I am not what happened to me, I am what I choose to become." She got multiple standing ovations.

"I made people cry. It was awesome," Beth said.

She remains active as a Certified Humor Professional, mentoring younger students in the program and spreading her message everywhere she goes.

But can any curriculum really teach people to be funny? And if so, how? I asked Morrison. She emphasizes that having a sense of humor is "very different from *being funny* or *telling jokes*." Rather, her message is that no matter how funny anyone is, they can always be better at being positive and optimistic, "being able to reformat negative things in your mind into something positive, no matter what it is. If you get cancer, that's something you cannot change. If you have people in your life who are what I call 'humor doomers' who suck the energy right out of you, though, that's something you can change. You can strategize to have less contact with those people."

In the more positive realm, she encourages students to keep

a journal of what made them laugh during the day. How did they take a situation that could've been negative and make it positive?

So you strengthen that muscle?

"Well, it's not a muscle," Morrison said. "It's neural connectivity in the brain."

The idea is that emotions and patterns of thought can be learned in just the same way that jokes can be memorized. With repeated "exercise," a neural pathway can be trained to be "stronger." Morrison is addressing the same kind of plasticity that allowed Beth to train the right half of her brain to control the right half of her body. In the case of humor, Morrison argues, if you grew up with parents and teachers who were prone to punishing and restricting playfulness, it could be less ingrained in your brain.

"So my work is on how you use humor to increase positive associations in your brain," she said. "I recommend people play every day. . . . I'm a swinger. I get on my bike every day, when the weather permits, and I go to the park, and I get on the swing."

Simple as the idea of purposeful play is, it's far from common, so Morrison believes her work is critical. So does Beth.

For the rest of his life, Fred Rogers called Beth on her birthday. Kathy asked him to be the commencement speaker at UConn in 1991. He agreed, on the condition that Beth help with writing the speech, which she did.

"The amazing thing about Beth is that she's happy," Seth Wohlberg said. "Most people who go through the hemispherectomy, they're not. That's something she seems to have figured out."

In Beth's case, apart from the removal of half of her brain, laughter was the only medicine. It's easy to dismiss the idea of a Certified Humor Professional, especially next to the more severe figure of the neurosurgeon. In most cultures, the latter is seen as doing the work that is important or real. But technical and intricate as it may be in practice, neurosurgery is only beginning to learn how to stop people from losing control of their bodies. The field is far from mastering the art of making life not just livable, but good.

In a 2009 biopic, Ben Carson was portrayed by Academy Award–winning actor Cuba Gooding Jr. and Carson was celebrated as a hero. Certified Humor Professionals like Mary Kay Morrison and Beth Usher don't receive the same acknowledgment, though they may deserve no less.

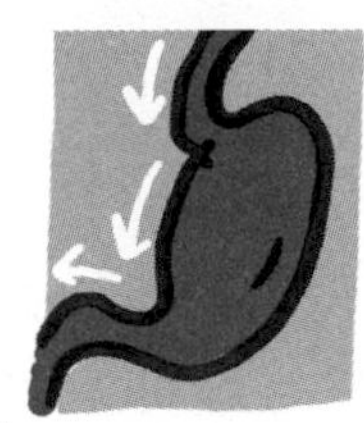

PART THREE

EATING

THE SUSTAINING PARTS

I felt like a junkie every day. Waking up starving, forcing myself to eat, you know, barfing it back up. Just imagine trying to eat your three meals a day, and just concentrating and just crying at times. Like [agghh]. I'm in pain all the time."

That's how Kurt Cobain described his six years of "constant pain" in his stomach. "They never figured it out," he said in the same MTV interview that aired in 1994, reclining before a red curtain, gaunt and dragging on cigarettes for thirty straight minutes. "Most gastrointestinal doctors don't know anything about the stomach diseases. . . . They just say, 'Oh, you have irritable bowel syndrome,'" he said, dismissing the expansiveness of the term. Cobain came to consider gastroenterology "a total scam." He tried numerous medications, prescribed and otherwise (including heroin), ultimately coming to the understanding that his condition wasn't "a specific stomach ailment. It doesn't have a name or anything. It wasn't a matter of finding out what disease I have. It's, you know, it's psychosomatic. It's part of my nervous system."

Cobain was ahead of his time in many ways, though he is rarely credited for his understanding of the enteric nervous system. Only in the decades after Cobain's death did we begin to understand how tightly intestinal functions are tied to emotional and cognitive functions of the brain.

Psychosomatic is a term from which patients tend to distance themselves. Many take it to mean "crazy"—that their symptoms are not "real." Cobain owned the term in a way true to its gravity and complexity. The system called the "gut-brain axis," which was used only in esoteric journal articles in 1994, is a means of two-way communication between the central nervous system (the brain and spinal cord) and the enteric nervous system (the constellations of nerves surrounding the bowels and stomach).

Only in a 2011 article in *Nature Neuroscience* did UCLA professor Emeran Mayer conclude that "gut-brain crosstalk" influences not only digestion, but also modulates our "motivation and higher cognitive functions, including intuitive decision making," and that "disturbances of this system have been implicated in a wide range of disorders, including functional and inflammatory gastrointestinal disorders, obesity and eating disorders."

This explains the onslaught of pop diet books written not by gastroenterologists but by neurologists. Added to that list now are microbiologists as we have begun to understand the role of the trillions of microbes (gut microbiota) that live inside our gastrointestinal systems in mediating these gut-brain interactions. This might be envisioned as a third element in a microbiota-gut-brain axis.

Disruption of the microbial ecosystem (termed *dysbiosis*) has been suggested to have a clear association with disorders of the central nervous system, such as autism, anxiety, and depression. The microbiota-gut-brain axis manifests through electrical signals across neurons, hormonal signals in the blood, and immune reactions throughout the body. A 2015 journal article from physicians at Sapienza University in Rome concluded that irritable bowel syndrome can be considered "an example of the disruption of these complex relationships."

At UCLA, Mayer, who is a professor of medicine, psychiatry and biobehavioral sciences, and physiology, is in a unique position to bridge the traditional boundaries of medical specialties. He founded a center for research in "Neurovisceral Sciences"—a new term—with the goal of understanding gut-brain dynamics. (He is specifically interested in people with chronic pain and irritable bowel syndrome, and why the conditions appear to act differently in males and females.)

Gastroenterology today still deals almost exclusively in the realm of the mechanical: cancers and ulcers and processes visible to the cameras at the ends of the scopes that can be dropped down our mouths and up our bowels. The field is ill equipped to deal with the more complex (and more common) causes of dysfunction, for which there is no single test or pathway.

As such, it might be easy to dismiss gastroenterology, as Cobain did. Among patients with irritable bowel syndrome, the relationship to the medical system can become antagonistic, as doctors report that they "couldn't find anything wrong." While this is often a simple failing of technology and the limits of our knowledge, to the patient it can read as an implication that she is lying, malingering, weak, or all of the above.

At the Tribeca Film Festival in 2015, Courtney Love said offhandedly that Cobain had Crohn's disease. While her knowledge of the situation is better than any I can presume to have, the stories conflict in a way that's illustrative. Because while the symptoms of Crohn's disease are similar to those of irritable bowel syndrome, Crohn's is a named disease precisely because doctors have long been able to recognize it (if unable to explain or cure it). Cobain would likely have known if this had been his diagnosis. He would probably have cited it by name, drawn as we are to a condition's name, whether or not it affects the prognosis.

It's telling, too, that people diagnosed with irritable bowel syndrome, which is functionally similar to Crohn's but less understood, are much more likely to suffer depression. In his 1994 MTV interview, Cobain seems to suggest that was the case for him. "I was in pain for so long that I didn't care if I was in a band. I didn't care if I was alive," he said flatly. "It had been going on and building up for so many years that I was suicidal. I just didn't want to live."

At the time, he reported that his symptoms were gone. But in his suicide note later that year, he wrote, "Thank you all from the pit of my burning, nauseous stomach."

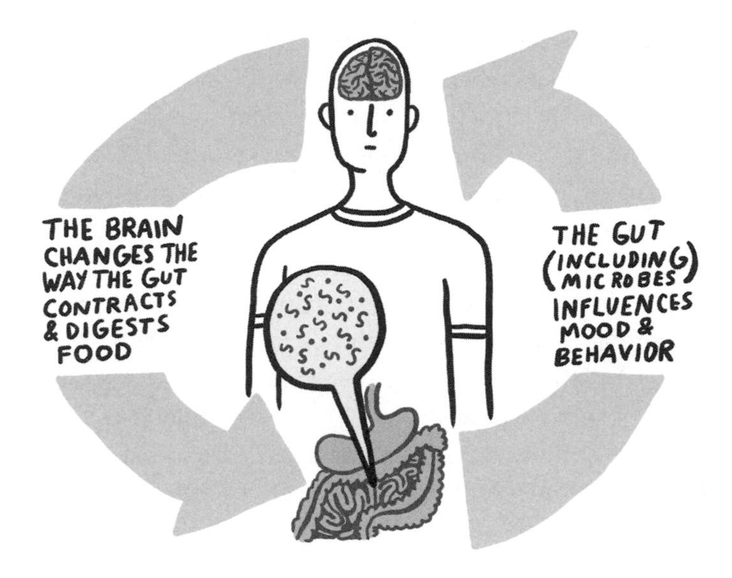

We are only beginning to understand the connection between our bowels and our brains. As we do, there's ever more focus on what we eat and its effects on health. For decades, food existed as part of a sort of tug-of-war between enjoyment and body weight. Now it's about more than that—everything from anxiety to acne to mental clarity to ADHD to cancer is being attributed, at least in some part, to our diets. Definitive answers are few and misinformation is rampant. But it's also easy to identify when equipped with a basic understanding of the human body.

Why do stomachs rumble?

Put your ear to anyone's abdomen, and within a few seconds you should hear rumbling and squeaking. If they ask you to remove your ear from their stomach, do so. Even in that brief moment, you should have caught the sound of the muscles in the walls of the stomach and intestines, which are almost constantly contracting. This functions to push food through the system, like a snake swallowing a mouse.

The sounds produced in the process are called borborygmi, the plural of borborygmus, and they're always there. They usually grow loud enough to hear only when air in the chamber allows the sound to resonate, like when you speak into an empty coffee cup and your voice sounds enormous, and you demand that everyone kneel before you.

In 2010, a British woman developed an extreme case of stomach rumbling that would not stop. Her doctors reported it in the medical journal *BMJ* as "intractable and refractory borborygmi." (Which translates to: "Her stomach just kept on rumbling, and nothing would stop it.") The rumbling stopped only when the woman lay down, but as soon as she sat up again, it would return. To attempt to figure out the cause, the doctors asked the woman to swallow some barium, which would coat the lining of her throat and stomach and appear bright white in a series of X-rays. The images showed a sort of luminous road map of her upper digestive tract. Unusually, the bottom of her rib cage angled inward over the middle of her stomach. When she inhaled, her ribs compressed her stomach. When she was lying down, gravity pulled her stomach toward her spine, and it escaped her concave ribs.

The doctors debated surgery to remove the offending ribs, but they weren't sure it would be worth the risks. The only way they could temporarily silence the borborygmi, they discovered, was by pressing on her left hypochondrium—the upper abdomen just below the ribs (*hypo* = below, *chondrium* = cartilage). This, incidentally, is where the term *hypochondria* comes from, as it was once believed that worry arose in the abdomen. This was laughed off when people discovered the central nervous system. But now, in light of the gut-brain-microbe axis, it turns out to be plausible.

In the patient's case, the hypochondrial pressure appeared to work by altering the position of the woman's stomach. So the five (male) doctors suggested she wear a tight-fitting corset.

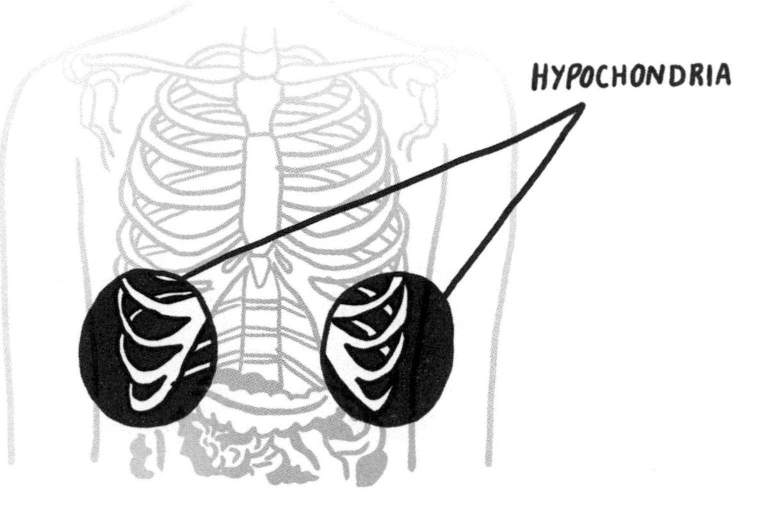

But they report in the medical journal that that did not work—perhaps, they suggested, because she did not wear it enough. And so they were left with no recourse but to translate her troubles into jargon. "Our patient continues to have troublesome, audible borborygmi," they wrote in the case report, "which in turn continues to cause social embarrassment."

Six years later, I contacted the physicians to see how this woman was doing. Her gastroenterologist Kieran Moriarty reported back with enthusiasm that "she is about 50% improved." Several weeks later, by email: "An update. Not done so well."

Why do I crave terrible food late at night?

One of the things that the University of Pennsylvania labs recently uncovered is how sleep deprivation causes people to gain weight. The head of chronobiology, David Dinges, and colleagues kept 198 experiment subjects in a lab and restricted them to four hours of sleep for five nights. A separate control group slept a decadent seven and a quarter hours. The subjects thought their performance was being measured (and it was), but the researchers also covertly measured their food intake and metabolic rates. In just five days, the sleep-deprived subjects gained, on average, a full kilogram.

"These late-night runs where people want to eat pizza and greasy foods, that's exactly what the brain wants," Dinges told me. "It's almost as though when you restrict your sleep, your brain says, 'I'm starving. I need fast-burning calories.'" Other studies have made similar findings.

Most of the extra calories went on board between 10 p.m. and 4 a.m. This is the time frame in which Taco Bell advocates, in ads, that people eat their "fourth meal." Food trucks and carts in the binge-drinking districts of cities do heavy business during these hours. I came to know the proprietor of a hot-dog cart in Los Angeles who worked outside some bars in Echo Park exclusively from midnight to 3 a.m. or so. And it's not just because drunkenness leads to hunger. Have you ever been drunk during the day? Probably not, but if you had, you'd notice that it rarely leads to the same kind of "insane" cravings for macaroni and cheese. In the middle of the night, on the other hand, you're likely to experience serious hunger even in sobriety, even though you don't "need" to eat. And after a night of sleep, you don't wake up starving, even though it may have been twelve hours since your last meal.

It's not just that the unrested subjects in the University of Pennsylvania study ate more food. With sleep deprivation, their resting metabolic rates dropped as well. Their bodies both took in more energy and burned less. Sleep and "metabolic disruptions" go hand in hand, Dinges believes. "Young people with fast resting metabolic rates who exercise a lot might be able to be sleep deprived without gaining weight," he said. "But as the years go by, you might gain weight at a rapid rate."

Colonoscopy: This is the best we can do?

As of this writing, it is customary for all people over a certain age to periodically have a camera on a mechanized tube as long as a person is tall inserted into them to detect and remove irregularities. The tube has a wire lasso at its tip, so that the doctor at the other end can resect anything suspicious growing from the intestinal wall—usually polyps that appear disposed to become a cancer. Colonoscopies are one of the few ways that we know to prevent cancers and detect them at stages early enough that they can be cured. That this invasive mechanism is among the best technology we have right now to detect and prevent cancer is a reminder that we have far to go.

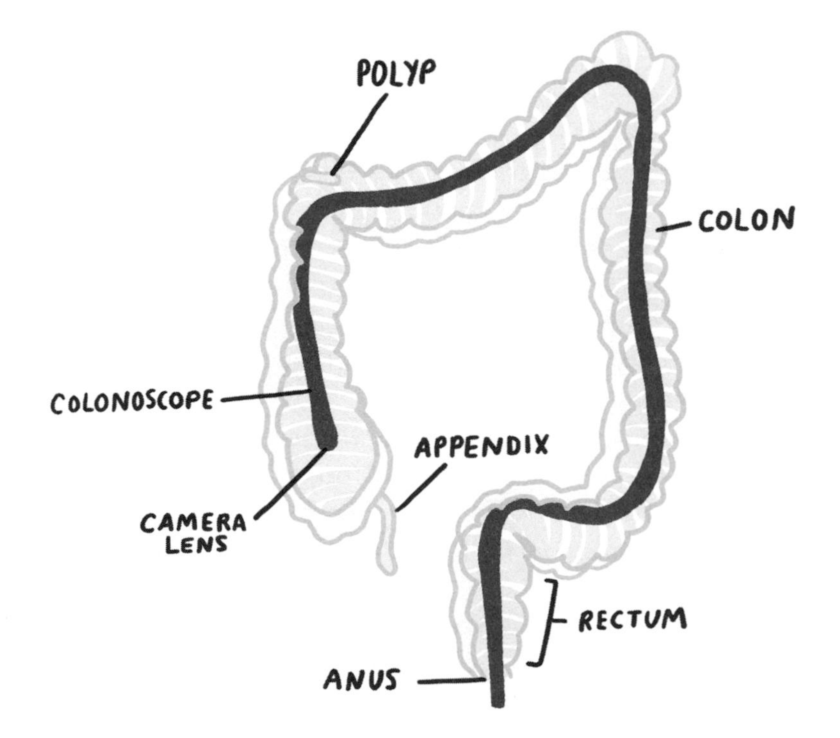

Is there any harm in taking a multivitamin? It makes me feel safe.

It came to the attention of British scientists in the nineteenth century that across the colonies in Southeast Asia, for no clear reason, people were losing feeling and control of their feet. Their legs would swell, and they had to swing their hips in order to walk. Their urine became bright, their chests tightened, and they lost balance, seized, and died. The people called it *beriberi,* signifying the waddling gait of a sheep, literally translating to "weak weak."

In 1803 in Sri Lanka (formerly Ceylon), Scottish army surgeon Thomas Christie attempted to cure the mystery condition by giving people vitamin C. (Technically, he gave people the fruits that were known to cure scurvy, as "vitamin C" had not yet been isolated as the active ingredient.) He was puzzled at why it didn't work, writing, "Giving 'acid fruits,' which I find of great value in cases of scurvy, has no effect in beriberi." He reasoned, presciently, that just because the compound was miraculous for scurvy did not mean that it would be of value in other circumstances.

Watching as patients continued to die en masse, Christie offered the alternative hypothesis that the disease was due to a toxin in water or food, a logical guess given the popularity of germ theory at the time. Accordingly, he gave the dying people laxatives to cleanse their bowels of whatever mysterious toxin was to blame.

(Many people take the same archaic approach to innumerable conditions today as a misguided measure in pursuit of general wellness. I get regular Groupon offers for "colon hydrotherapy." Dear Groupon: Collective purchasing can't make this a good deal.)

Even after the people were colonically purified, predictably, the suffering continued. It appeared in ever more countries. Epidemiologists mapped the outbreaks and noticed that the disease was common among populations that ate a lot of white rice. And when people moved away from diets that relied heavily on white rice, the people recovered. It seemed clear that rice was the cause. As with scurvy, the effect appeared miraculous; people stopped seizing and recovered completely, sometimes within hours. For decades, though, no one knew why. The mindset that diseases were caused by discrete entities—microbes or toxins—blinded many to their actual causes.

Obsessed with trying to find the compound responsible for beriberi, few considered the alternative. Beriberi did turn out to be caused by white rice, essentially—but not by a toxin in it, or an allergy or "sensitivity" to it. Christie had been correct at the outset in 1803: The weakness syndrome was indeed caused by something that white rice did *not* contain.

As technology allowed milling rice and removal of the husks, the diets of large populations of people fundamentally changed. In switching from eating whole brown rice to eating only the endosperm (known as white rice), people became deficient in a compound found in the bran of the rice. When that compound was identified as thiamine pyrophosphate, it became the first compound dubbed a *vitamin*—a discrete chemical in food that prevents disease. It plays a part in several basic bodily processes, primarily in reactions that metabolize carbohydrates and amino acids. But the compound fundamentally transformed how we think about our bodies today.

The suffix *-amine* means that a compound contains nitrogen, specifically in a basic form with a pair of lone electrons. Because thiamine was the first named chemical that needed to be eaten—lest we die—the Polish chemist Casimir Funk combined "amine" with "vital" (giving life), coining "vit-amine," which became "vitamin." The term first appeared in publication in his 1912 paper "The Etiology of the Deficiency Diseases," which combined the four known conditions that appeared to be caused by a lack of *something* in the diet: beriberi, scurvy, pellagra, and rickets.

Even though thiamine was the only "vitamin" compound that had been identified at the time, Funk presumed that the other "deficiency diseases" would be explained by similar vitamins: "We will speak of a beriberi or scurvy vit-amine, which means a substance preventing this special disease." It turned out that the pellagra vit-amine was not an amine at all, but we still call niacin a vitamin. The scurvy vit-amine was not an amine either—it was ascorbic acid—but we call it vitamin C. The rickets vit-amine was a pre-hormone, but we call it vitamin D.

There are now thirteen chemicals whose absence can cause deficiency diseases and thus carry the name vitamin. They are united by no common structure or function, only by the general notion that we cannot live without them. In her thorough history *Vitamania,* Catherine Price recounts that many scientists who adopted the use of "vitamin" considered it a temporary term that would become obsolete once they figured out what these compounds really were.

But that never happened. The meaningless name persists out of tradition. The term "vitamania" itself was coined during the vitamin craze of the 1950s, which has never truly passed—partly because the name is of enormous value to an enormous industry that sells a seductive concept. Had Funk never invented the word "vitamin," thiamine pyrophosphate might have continued to be called "anti-beriberi compound." And vitamin C might still today be called "antiscorbutic compound." These are appropriate names for specific compounds that, when absent in a body, lead to specific diseases. Would people buy foods today because they were "fortified" with "antiscurvy factor"? Would they believe that thirty times the necessary daily amount of antiscurvy factor is better than one? Maybe, but it's easier to justify when these things are said to be life-giving *vitamins.* (The exact same thing has begun to happen with *probiotics.*)

Most of the thirteen vitamins are coenzymes, which help specific enzymes affect specific chemical reactions. Each person may have a different need. Some postmenopausal women, especially in places where sunlight exposure is rare, may benefit from

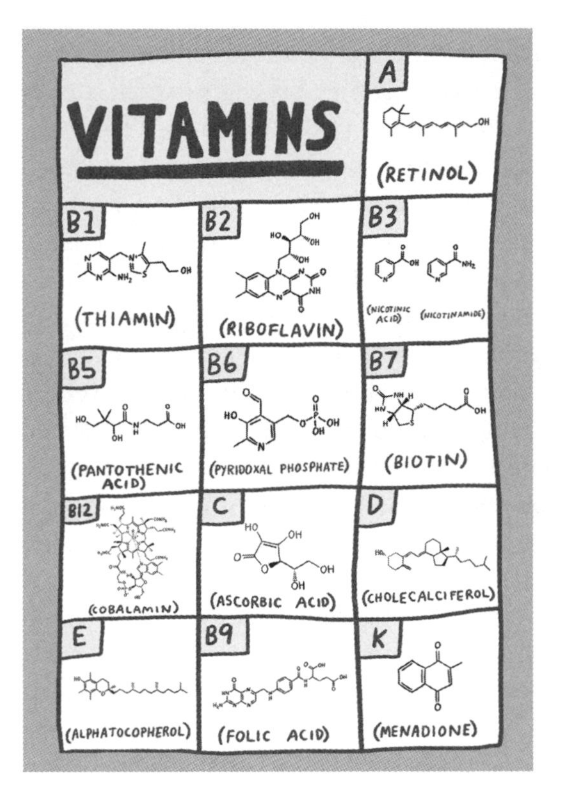

calcium and vitamin D supplements to mitigate osteoporosis. The American Academy of Pediatrics likewise recommends that breast-fed infants get daily supplements of a tiny dose (400 IU) of vitamin D.

The clearest case for taking a vitamin supplement is folic acid (or vitamin B_9) in pregnant women. The chemical is critical in signaling an embryo's neural tube to close during the first weeks of pregnancy. When the tube does not completely close, an infant can be born with everything from a cleft lip to a malformed spine that leaves the spinal cord entirely exposed to the world, called spina bifida. Consequently, the Centers for Disease Control recommend that everyone who might potentially become pregnant take 400 micrograms of folate every day. Even though this is a large dose compared to what most people need, it would still fit on the head of a pin and is easily attained in a diet based in whole plants. The need for taking a supplement is only because so many people today rely on foods stripped of micronutrients.

But no other supplement has such a clear case. Even putting aside the dangers of overdose, there are always risks to introducing "supplements" into your system. In one egregious case, the brand Purity First's line of "Healthy Life Chemistry" supplements were found to contain anabolic steroids. Purity First's "Vitamin B-50" (there are six recognized B vitamins, of which B-50 is not one) contained steroids called methasterone and dimethazine. Females reported unusual hair growth and missed menstruation, and males reported impotence. After twenty-nine people filed formal complaints about the drug, the FDA asked the company to please recall their product. Purity First declined to do so. When the FDA threatened to sue, it eventually complied.

Purity First Health Products is still in business in Farmingdale, New York. It is one of thousands of supplement manufacturers that is, somehow, still successfully selling purity.

Still, these vitamin compounds came to be perceived as panaceas. As indicators of virtue, even. When the entertainer Hulk Hogan was suing the website Gawker for character defamation in 2016 (after it published a tape of him having sex with a friend's wife), he claimed that his "impeccable" image had been shattered. That image, he said at the time, was "the all American hero, you know, the training, prayers, and vitamins." (The suit was backed by tech visionary Peter Thiel.)

That image also included more than a decade of anabolic steroid use, according to Hogan's own 1994 testimony. Risky as that was, the heroic consumption of high doses of vitamins appears to be generally harmless for most people. Of the thirteen vitamin compounds, overdose is a concern primarily for the four that dissolve into the fat in our bodies and accumulate: A, D, E, and K. Because these fat-soluble vitamins build up in our tissues over time, usually insidiously, the effects might not be evident for years. The other known vitamins are soluble in water, so we can overdose and they should safely pass through our kidneys and into the toilet. You can watch this happen after taking most multivitamin products, which tend to turn people's urine ultra-yellow as they overdose on riboflavin (from the Latin *flavus,* yellow), also known as vitamin B_2, and the kidney attempts to restore balance by excreting the excess. Some multivitamins have nearly one hundred times the recommended daily allowance of riboflavin. This is as close as most people come to watching their money go down the toilet.

Medical experts have repeatedly and definitively advised that people not consume vitamin supplements as though more were better: Vitamin supplements are bioactive substances, like drugs, and should be afforded similar skepticism. And it is especially difficult to predict the effects of products that combine multiple chemicals at once, known as multivitamins, and presume that the combination will have no bearing on their effect.

Certain groups of people might benefit from bundles of vitamins, like those with anorexia, or those in the wake of trauma that prevents consumption of food by mouth. Children recently removed from abusive households are at high risk for being broadly nutrient-starved. But as a rule, the answer about multivitamins is a clear and exasperated *no*.

In 2006, a group of researchers with the U.S. Department of Health and Human Services published a comprehensive review of all the studies that appeared to suggest potential impacts of multivitamins on health. Among people who eat even somewhat reasonable food, multivitamin supplements did not reduce the risk of *any* chronic disease (though there was no clear evidence of harm, "except for skin yellowing by beta carotene"). The U.S. Preventive Services Task Force did a similar review and found the evidence for using multivitamin products to prevent cancer or cardiovascular disease was "insufficient." More emphatically, the World Cancer Research Fund International and the American Institute for Cancer Research have recommended against dietary supplements as a matter of cancer prevention "because of the unpredictability of potential benefits and risks, as well as the possibility of unexpected adverse events."

These studies and expert recommendations against taking multivitamin products have been reported repeatedly, in innumerable widely read magazines and newspapers, and yet around a third of Americans continue to take them.

Unlike smoking less or exercising more, forgoing multivitamins is a health recommendation that involves no effort. It is not dependent on having access to a gym or a physician—or any resources at all. It does not cost anything; it only saves a person money. It is one area of public health where experts recommend we take *less* effort on something, and yet people persist.

Why does everyone have bad breath?

Gary Borisy left his career in midlife to do what he loved: study the bacteria in our mouths. He was a biophysicist in 2013 when he decided to leave the field for oral microbiology, the perfect place to study a dynamic ecosystem. Because our mouths are, in his words, "open sewers."

The consequences of understanding that ecosystem are practical not just in preventing teeth from rotting and falling out, but because, Borisy believes, "bad breath is probably what we worry about more than anything in terms of frequency of concerns. You go into any social situation and you check to see whether your breath is okay."

I felt self-conscious then about the fact that I don't check my breath very often (never). Do other people? I'm not sure I'd know how.

Borisy believes that when it comes to mouth odors, the tongue is where the magic happens (magic in this case meaning halitosis). Much of the smell that emanates from our mouths comes from bacteria on the tongue that are making volatile sulfhydryl compounds, which give breath that classic garbage smell. Colonies that form plaque on our teeth also produce some things that are foul-smelling, Borisy explained to me, but most of it comes from the tongue.

But why, evolutionarily, would we have these sewer mouths?

"There's literature out there that's not widely known, but it should be—you're an MD, right? Have you encountered what's called the *entero-salivary circulation*?"

Nooo. (*Entero* meaning "bowels," and *salivary* meaning "of or pertaining to saliva.") Borisy explained that there are bacteria in the mouth that convert nitrate in our diets into nitrite, which then goes into the stomach and gets converted to nitric oxide. This seems to be part of a homeostatic mechanism, which leads to lower blood pressure.

"When we take mouthwash and kill the bugs in our mouth, we're killing off some of these nitrate-producing bacteria," he explained, "so we may have sweet-smelling breath but increased risk of dying from stroke."

Oh my God—additional studies needed. It's tempting to hear that kind of early hypothesis and want to stop brushing your teeth. But at this point, it's just an example of the possible, plausible explanations for why we harbor so much mouth bacteria. The point is that the ecosystems in our mouths are not isolated from the health of the rest of our bodies.

"I'm *not* saying that the ones that produce bad breath are helpful," Borisy clarified. "Those are down deeper in the tongue. But the nitrite producers at the surface of the tongue do give us benefit. Of course, we benefit *them* by providing a nice surface to live—a very clear mutualism on the tongue."

The idea of mutually beneficial relationships may explain why we keep so many sulfur-producing bacteria in our mouths, too. They may not benefit us directly, but they may benefit the beneficial nitrite-producing bacteria. Stoic as they appear, teeth are dynamic scenes.

The white enamel that protects the nerve-filled root is

destroyed by acid erosion. We call this "decay," but decay actually refers to the fact that bacteria in our mouths ingest sugar and ferment it. When you're making beer, say, or synthesizing vitamin C for your supplement business, that fermentation is a good thing. When it's happening in your mouth, less so, because the process releases lactic acid. That dissolves calcium in the enamel—especially in the little crevices where bacteria hang.

It's through understanding this process that scientists could someday free people from brushing their teeth at all—if there were another, more precise way to manage these ecosystems. Part of the reason Borisy switched fields to study plaque was because he saw a gaping hole in knowledge across the board.

"I saw this revolution in DNA sequencing being applied to the microbiome," he recalls, "but there is a missing layer." That layer was the structure of the microbial ecosystems. To sample your gut bacteria, for example, a lab has to grind up a fecal sample in order to get the DNA sequences. That can tell you what bacteria are inside you, but it tells you nothing about their relationship to one another, the structure of the colonies. It's rather like trying to understand a person when all you have is their disassembled brain.

In 2016, Borisy and colleagues published the first 3-D fluorescent images of the bacterial colonies in tooth plaque. They wanted to see exactly what species were where, and how that relationship seems to work.

We produce 1.5 liters of saliva a day, so everything in the mouth would get flushed into our stomachs were it not adhering to something. These streptococci adhere to bacteria called *Corynebacterium,* which affixes itself to the white enamel of our teeth. This bacteria's function seems to be simply creating a frame for the colonies that build plaque. (Which can exist because *Corynebacterium* produce an enzyme that destroys the bacteria-killing peroxide produced by strep.) It is this sticky skeletal layer that makes plaque hard and difficult to remove. This is

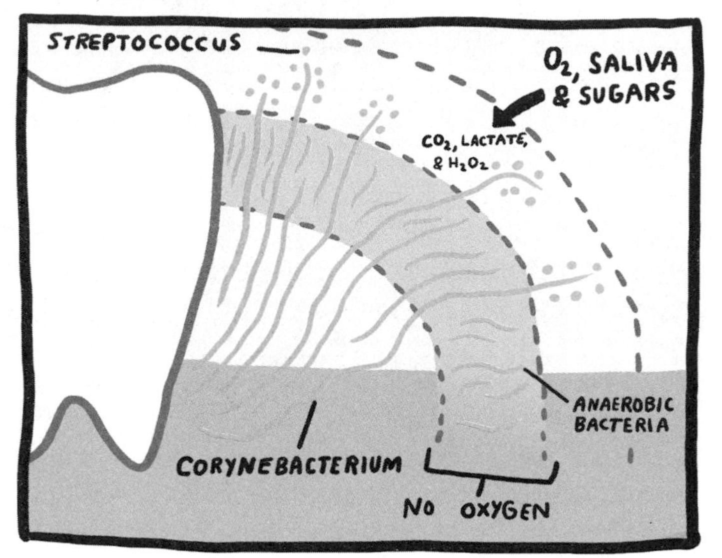

the reason why hygienists must scrape at our teeth with metal instruments, using degrees of force that seem, at best, undue. (The scraping sound of metal against enamel can make people physically ill, especially if they have misophonia.)

The acid-producing streptococci also secrete hydrogen peroxide, which kills other bacteria. So even though the acid erodes our teeth, the strep may be beneficial because they subdue other "bad" infection-causing bacteria. The strep produce carbon dioxide, too, which creates ideal environments for some beneficial species (like *Capnocytophaga* and *Fusobacterium*). To Borisy, "if there's one biology lesson that comes through, it's that function is connected to structure."

So it may be that we have bad breath and eroded teeth, but only because it's better than having strokes and abscesses. Our bodies are the way they are because it is better than the way they are not.

The images of the plaques are like elaborate coral reefs rendered on a late-1990s Windows screen saver. They make me never want to brush my teeth again and disrupt their beauty.

Borisy cautioned me against this. "I think what we should be aware of is that there is a complex ecosystem in our mouth," he told me. "We ought to think twice before we disrupt it."

You're not suggesting people cut back on brushing?

"I'm not suggesting they cut back at all," he said. "I'm just pointing out that [elaborate microbial ecosystems] are there, and they might be there for a reason."

Carbs or fat, which is worse?

Private as most of us are about what happens inside our bowels, they are our largest interface with the outside world. The average person in the United States eats 1,996 pounds of food every year. What we eat may be the most consequential decision we make in our lives not just in terms of our health but, collectively, in terms of the global economy and the environment.

As the Internet demands more and more words for consumption, hundreds of food articles seem to be published daily. As that number grows, it becomes ever more difficult to stand out without a mind-blowing story. There is incentive to exaggerate the importance of the latest study, to suggest that it is changing the game. (Would people read a story called "How to Eat in a Way That Will Make You Statistically Unlikely to Gain Weight: Basically the Same Fundamental Principles You've Already Heard"?) The subtle effects of each haphazard story and fad add up to a serious problem for public health.

It is because of this problem that I came to Boston on a warm November weekend in 2015. I was included in a meeting of food-world-famous nutrition scientists, organized by David Katz, director of the Yale-Griffin Prevention Research Center, and Walter Willett, chair of the Department of Nutrition at Harvard T.H. Chan School of Public Health. The name tags hanging from red lanyards around the necks of the mostly older white men in modest suits throughout the Hyatt Regency Boston Harbor conference room represented a surreal who's who of public nutrition. They were twenty-five people who have put themselves most forthrightly into the public conversation. Their goal was to undo the perception that nutrition science is chaos, and to unite around some common principles about food and health that would be useful to the world.

There was T. Colin Campbell, the author of *The China Study,* which has become the human-health basis for much of the modern vegan movement. Campbell grew up on a dairy farm, and went into his nutrition PhD training in 1958 with an eye to figuring out exactly why milk was *such* a superior food. Now a white-haired professor emeritus at Cornell, he believes milk is a carcinogen—and despite multiple bestselling books, he remained just as adamant in telling me so. There was also Stanley Boyd Eaton, a retired medical anthropologist and radiologist at Emory University who was among the creators of the modern "Paleo diet," and Tom Kelly, the chief sustainability officer at the University of New Hampshire's Sustainability Institute. All the way from Athens, Greece, came Antonia Trichopoulou, director of the WHO Collaborating Centre for Nutrition. At the University of Athens, she led the "Mediterranean Diet" into worldwide recognition as a healthy way of eating. (To her, *the* healthy way.) There was also Dariush Mozaffarian, dean of the Tufts University Friedman School of Nutrition Science and Policy; the inventor of the glycemic index, David Jenkins; the Harvard nutrition professors David Ludwig, Frank Hu, Meir Stampfer, and Eric Rimm; and the legendary Dean Ornish, professor of medicine at University of California, San Francisco, who seemed to command a degree of celebrity even within this high-status crowd.

Throughout a long day of lectures, each made their succinct case for the way of eating they believed was best for health—the ways on which they had based their long careers as researchers, and many as speakers and authors. In the evening, they sat down around an enormous table to "find common ground."

Co-organizer David Katz seemed to think the task would be easier than it was—that they would have a nice document by the end of the meal. He stood at an easel that held a blank white flip board. I was the only journalist in the room, and I agreed not to print any specific quotations from people so that everyone could brainstorm freely, but it went like this. Someone started with the proposition that should be least controversial:

Can we say that everyone should eat vegetables?

Most nodded. Then someone said, well, what kinds of vegetables?

Cooked or raw?

Right, I was just thinking that, because we can't have people eating only white potatoes. Those are pure starch.

Are French fries and ketchup vegetables?

I think people know that when we say vegetables, we don't mean to eat purely French fries.

Do they?

The federal school lunch program says they are.

[Lots of talking over one another.]

Well, what if we say to eat a variety of different-colored vegetables?

I don't think that's supported by evidence.

Does it have to be?

YES.

They could still cover the colorful vegetables in salt, and that wouldn't be good.

Or deep-fry them.
Should we say raw?
No! We can't neglect the importance of flavor!
And cultural tradition.
What about seasonality? We can't have everyone eating avocados all year round.

After the first hour, we had arrived at no consensus on whether it could be said that people should eat vegetables.

Over the next four hours, it became clear why this sort of consensus statement does not exist. Every one of the twenty-five scientists in the room did, very clearly, agree that people should eat vegetables. And fruits, nuts, seeds, and legumes. They all agreed that this should be the basis of everyone's diets. There should be variety, and there should not be excessive "processing" of the foods. The devil was in how to say this. They stayed until midnight that day in Boston, trying to figure it out.

Ultimately the main point they agreed on was that serious problems result from their perceived discord. When people sense the absence of a single established consensus on nutrition, this invites them to see every diet trend as equally valid. It gives credence to whatever the latest news story suggests, or whatever the Kardashians are doing, or to whoever is selling the latest book about how carbs/fat/gluten are "toxic." As Robert Proctor, the Stanford professor of agnotology, notes, the "experts disagree" strategy is key to cultivating ignorance. It is a tactic of demagogues to make people believe that no one knows anything—so you might as well believe their absurd idea.

Experts can and will continue to disagree on how to interpret bodies of evidence; this is a foundational element of science. It means the process is working like it's supposed to. But that doesn't mean there isn't consensus about many of the tenets of nutrition.

It was in hopes of quelling this very discord that all of these nutrition experts came to Boston. There they agreed and put in writing that, yes, eating mostly plants, ideally in their whole forms, is clearly advisable for human health, both as individuals and as a collective human body.

They agreed that food is medicine; it is the lowest-hanging fruit in public health, in a world where most people die of preventable cardiovascular disease caused largely by eating poorly. But nutrition is about much more than preventing discrete diseases. Every compound that we apply to our bodies, inside and out, has effects. Each dietary decision is small, but after years of eating—several times a day, as most people do—the effect on our health and well-being is immeasurable.

They also agreed that while they speak and write often about carbohydrates, proteins, and fats, it is not advisable as a practical rule to think about eating in reductionist terms—that is, to focus on one compound in food and then demonize or deify it, tempting as that psychology is.

The idea of thinking about food as carbohydrate, fat, and protein stems from an 1834 book by William Prout, *Chemistry, Meteorology, and the Function of Digestion,* in which he posited that food is composed of three energy-containing "staminal principles." True as this was, it was akin to saying that the solar system is composed of planets and a sun. And yet the carbs/fat/protein breakdown is still, two centuries later, the way many people think about nutrition today—the basis, even, for the nutrition labels in many countries.

But the 1830s concept of carbs, fat, and protein was as simplistic as most of the science of the era. The three groups break down much further, into myriad types of each compound. The plants that we know to sustain life contain not only the compounds we call vitamins and minerals but also other phytochemicals whose import we are only beginning to understand. And even when we do know everything and can administer it in a formula that is theoretically identical to what is in a plant, we will be violating the fundamental tenet of functional biology: that there is importance to form. That the whole is greater than its parts in sum. In hospitals, when a person is too sick to eat food by mouth or gastrointestinal tube, the best possible nutrient concoction (known as total parenteral nutrition) can be given directly into their veins. Even under the strictest of meticulous calculation and monitoring, though, this will only keep a person alive for a matter of months before their livers fail and their gut bacteria are decimated.

So emerges the simple recommendation for human health, as individuals and as populations, sustainable and attainable and customizable to myriad cultural traditions and loves of flavor and budgetary considerations: a whole-plant-based diet.

What is gluten?

On the twentieth anniversary of Kurt Cobain's death, an addiction specialist in California named Morris Mesler reflected on the artist's gastrointestinal torment. "It might merely have been a question of slightly altering his diet," Mesler said, suggesting that Cobain may have had "lactose or gluten intolerance."

On Reddit and other forums, slightly less qualified people (anonymous commenters) continue to speculate that Cobain had celiac disease. One gluten-free blogger waxed recently, "I think a gluten-free diet would have helped him feel better and possibly saved his life."

Cobain's list of exposures included inordinate alcohol and tobacco, heroin, and inhaled solvents. He carried a diagnosis of bipolar disorder and the trappings thereof. Yet today it has apparently become reasonable to point fingers at gluten.

Within wheat, rye, and barley are two proteins, gliadin and glutenin. When flour is mixed with water, those proteins combine and form another protein, gluten, which has a more elaborate matrix structure than either of the two individually. It allows dough to be viscous but also elastic—cohesive but also malleable. These are good qualities in baking, as in people.

Gluten allowed *Homo sapiens* to create bread, which became

a staple of diets around the world as the species thrived and spread. But gluten itself is far less interesting than our new, tortured relationship to it—and the money and psychology on which that relationship subsists. Gluten has come to offer identity, resilience, and structure not just in baked goods, but in life.

With stakes so high, the critical first step in every discussion of gluten is to acknowledge that the symptoms a person is experiencing are reality. People who misrepresent themselves will be integral to the story of gluten, too, but suffering people who believe themselves sensitive to gluten should not be lumped among them. The second step is to look at what we know about this protein, and about our bodies, and the likelihood that gluten is the cause of so much real suffering.

In the Nazi-occupied Netherlands, November of 1944 to May of 1945 would become known as the Dutch Hunger Winter. Nineteen thousand people died of starvation. As if by some isolated miracle amid the scourge, when flour became scarce, some people who had long ailed with a vague condition known simply as

HOW GLUTEN WORKS

FLOUR
WATER
GLUTENIN & GLIADIN → GLUTEN

celiac disease ("abdominal" disease) recovered. In the following decades, British researchers determined that the offending agent was not flour itself, but the protein gluten.

About 1 percent of people are now known to have celiac disease. It is the quintessential "autoimmune disease," by the traditional understanding of that term. Our bodies mistake *self* for *other*. In this case, specific antibodies (called tissue transglutaminase) destroy the lining of the small intestine whenever gluten is present. This leads to intense indigestion, and malnutrition in severe cases. As the bowel wall is destroyed, so follows derangement of the gut-brain-microbe axis. Some people report headaches, seizures, numbness in the fingers, and depression. The disease can disrupt every aspect of life, causing short stature to anemia to miscarriage.

A clear diagnostic test can tell us if a person has these antibodies. For people with them, the diagnosis is celiac disease, and the only known cure is avoiding gluten completely and categorically. And unlike most other treatments for disease, withholding gluten is effective always. It is one of the clearest-cut situations in medicine, a rare binary: Either you have antibodies that destroy your bowel wall when you eat gluten, or you do not.

In this way, celiac disease is like scurvy or beriberi. This may be partly why gluten has become, in a psychological inverse to that applied to vitamins, subject to the same kind of monotonic absolutism and extremism. A compound is bad for people in one (very clear, limited) context, so it must be bad in many contexts.

Just as high doses of ascorbic acid (vitamin C) have no discernible effect on influenza, avoiding gluten has no discernible impact in people without celiac disease. This does not mean such mechanisms may not be discovered in the future, because anything is possible; but there are many more plausible things to be undertaken for health benefits than the wanton avoidance of grains. Yet as with many of the areas in medicine where shades of gray are least relevant, people are most likely to see them.

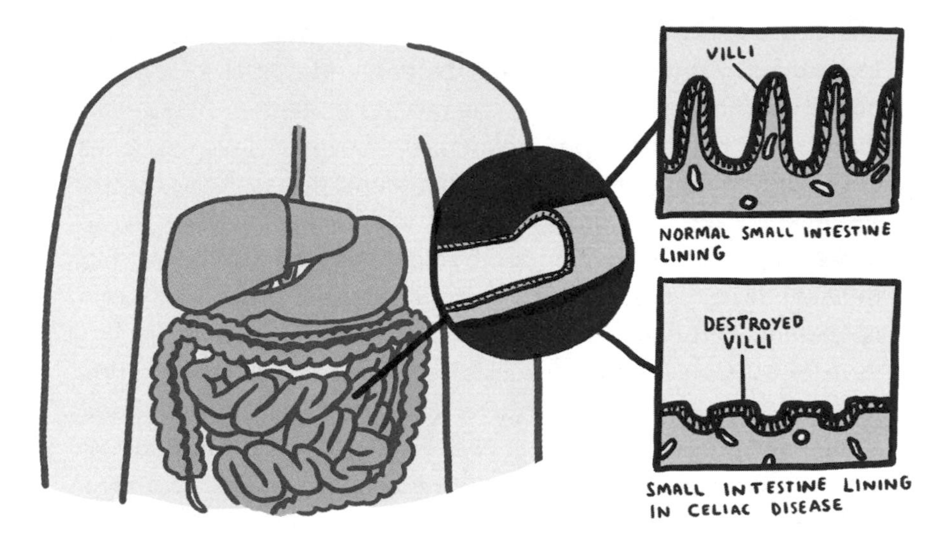

"Gluten free" is now, worldwide, the most Googled approach to eating. Between 2012 and 2015 alone, sales of products labeled "gluten free" doubled (from $11.5 billion to more than $23 billion). There are entire gluten-free bakeries, gluten-free dog foods, and gluten-free cosmetics. That this bubble of fear is growing around a plant protein is a case study in effective marketing and mob mentality, and a testament to how badly we want to be healthy. It's enough to cause a reasonable person to wonder, "Wait, should I be avoiding gluten?" and then, with a shrug, think, "Why not?"

In a nondescript late-twentieth-century brick commercial office complex in Phoenix sits Cyrex Laboratories. I learned of it in 2016, when the company's publicist wrote me an email with the subject line "How to Know If You Should Be Gluten-Free." The promise therein was a "simple blood test that accurately identifies gluten reactivity." This test was marketed not for celiac disease, but toward the millions of people who believe they may have some kind of sensitivity to gluten that hasn't as yet been provable.

It immediately caught my eye because that product should, if it worked, be a tremendous advancement in gastroenterology, and help retire the fad. *Would you benefit from avoiding gluten? Take this simple test.* Such a product is a goal of scientists at academic institutions around the world.

The publicist directed my myriad questions about this test to an expert who "can speak to Cyrex's test [and] the symptoms someone may experience that signal if testing should be done." She called him "Dr. Chad Larson." The quotes are mine. In many countries, the word *doctor* implies an MD, PhD, or sometimes a DO degree. Larson has none of these, but is part of a long-simmering, now-exploding movement of doctors whose profession is predicated on the fluidity of language. Emoji are words now, and "doctor" can mean many things. Larson uses the title "Dr." because he holds an "ND" degree. That stands for doctor of naturopathic medicine, which is not to be confused with an "MD," or doctor of medicine. (Or, actually, is it?)

The subject of Larson's degree cuts to the heart of the gluten sensitivity epidemic. Gluten sensitivity divides us into those who are given to see a black-and-white world—where concepts or compounds are wholly bad or wholly good—and those who are given to inhabiting nuance and uncertainty. Gluten sensitivity is less about our relationship to gluten than it is our relationship to knowledge.

An MD (*Medicinae Doctor,* "teacher of medicine") is the traditional degree that has long defined doctorhood, earned in the United States from one of 141 accredited four-year graduate schools by people who have completed premedical studies and the MCAT exam. Since 1876, the schools that educate medical doctors have been overseen by an organization called the Association of American Medical Colleges (AAMC). These graduate schools are required to train students in "advances in medical knowledge, therapies, and technologies to prevent disease, alleviate suffering, and improve quality of life," along with teaching "concern for compassion, quality, safety, efficacy, accountability, affordability, professionalism, and the public."

Larson and other naturopathic doctors espouse an "alternative" to this approach. His ND degree comes from a school called Southwest College of Naturopathic Medicine, which is one of seven institutions that have sprung up in a new consortium, outside the purview of the AAMC. They did, however, establish their own accrediting body in 2001, calling it the Association of Accredited Naturopathic Medical Colleges (AANMC).

The AANMC's website is a near replica of the AAMC's. It even similarly uses as its symbol the caduceus, long-standing emblem of the medical profession, the Greek messenger god Hermes' winged staff ensconced by two snakes. The only difference in this case is that in the AANMC's caduceus, the intertwined snakes' heads are replaced by two leaves. The leaf-snakes are, at first glance, a benign homage to nature. But a leaf-headed snake is only sort of natural.

Naturopathy seems to have arisen like a rebound relationship. Its practitioners have positioned themselves as an "alternative" to a medical system that is (indeed) inefficient in many ways. MDs often fail to consider social contexts of health, and medical science can explain only a small fraction of known human pathology and how to effectively treat it. But by positioning itself as antithetical to that, the AANMC has avoided saying what it actually *is*.

Central to naturopathy is the practice of "prolotherapy," for example, where a practitioner will inject a dextrose solution (literally sugar water) into a person's joint, spine, or wherever they are experiencing pain. The practice is not just unsubstantiated, it doesn't even offer a plausible mechanism for how it could be effective. Yet in certain circles this lack of evidence is a virtue. When something has been tested for effectiveness and quality, it gets adopted by the establishment and loses its "alternative" edge. Then it becomes simply medicine.

For now, Chad Larson is convinced that gluten is the cause of inordinate suffering.

"If a person has a chronic issue, you can kind of fill in the blank, whether it's migraines, irritable bowel syndrome, multiple joint pain, chronic low back pain, thyroid imbalance," he told me. "If anyone has a chronic problem and gluten is part of their diet, it's at the top of the list of things I'm going to check for."

Is there a case where you wouldn't think that gluten might be the cause of someone's symptoms?

"I can't think of one."

It is worth noting that Larson is a paid consultant to Cyrex Laboratories, the company whose publicist urged me to speak with him. To ride an air of antiestablishment sentiment so righteously that it distracts people from one's *other* establishment is a feat of agnotology.

While celiac disease had been known since the 1940s, "non-celiac gluten sensitivity" (NCGS) was coined in 2012 by Harvard University professor of medicine Alessio Fasano and colleagues. And as the person who named and defined the condition, Fasano explained that there's no evidence behind Cyrex's gluten tests.

"They've not been validated," he told me. "I don't know on what kind of assumption they believe these are good biomarkers for gluten sensitivity."

The blood tests sold by naturopaths like Larson are not even recommended by the nonprofit Celiac Disease Foundation, which espouses a similarly all-encompassing understanding of gluten sensitivity. ("People with gluten sensitivity can experience symptoms such as 'foggy mind,' depression, ADHD-like behavior, abdominal pain, bloating, diarrhea, constipation, headaches, bone or joint pain, and chronic fatigue when they have gluten in their diet, but other symptoms are also possible.") Yet the foundation states clearly on its site, "There is currently no blood test for gluten sensitivity."

When you talk about "immune responses," Fasano explains, they come in three varieties. First is a classic allergy (as to peanuts, shellfish, or wheat). Allergies are easily detectable, marked by antibodies that cause inflammation while attacking, say, peanut dust, producing symptoms that are a sort of collateral damage. The second variety is an autoimmune reaction: A food causes a response wherein the immune system directly "attacks" other cells in the body. Celiac disease is the classic example. Gluten causes a person's body to attack its own bowel cells, obliterating them. In the absence of gluten, the immune system doesn't attack those cells, and the person is fine. Finally, there are what's known as food sensitivities.

"Here," he said, "we are in an almost unexplored territory."

Fasano created the name "non-celiac gluten sensitivity" in a way that has more in common with diseases classically attributed to the mind than the body. As the difference between conditions of the body and mind has faded, the distinction between such conditions is rather in how they are defined. Psychiatric diseases are defined by observing a set of symptoms and then giving a name to that set of symptoms. This runs in contrast to the diseases that fall to other doctors, where a condition is determined by a quantifiably abnormal biological process. In line with the standard psychiatric model, Fasano, like so many celiac disease specialists around the world, saw patients who did not have discrete biological conditions such as celiac disease or wheat allergies, but did say they had symptoms that got better when they cut back on gluten.

There is, though, good reason not to blindly pin so much illness on gluten. Medicine has a thorough history of presuming to know the cause of disease and being mistaken, and in so doing, overlooking the actual cause. It happened when we believed that cholera was spread through the air, and when we believed beriberi was caused by a toxin in rice, and it seems to be happening now with gluten.

The term "gluten sensitivity" is itself misunderstood. For people whose bodies are demonstrably debilitated by celiac disease, the name can be an affront. Mary Schluckebier, who leads the Celiac Support Association, typified the animosity in one interview: "A patient goes to the doctor, and they want a diagnosis. 'Don't tell me I don't have this—give me a name for it.' So doctors came up with a name for it. I think it was a way to appease impatient patients. I don't know how to say that nicely."

It may be that this frankness and honesty is actually the nicest way to say it. Schluckebier must walk a fine line, lest people get the mistaken impression that they are being dismissed as liars or malingerers.

"This is a work in progress," Fasano is quick to say of NCGS, exercising caution in defending his diagnosis. "We're at the very beginning. We still don't know who gets this, what the natural history of the disease is, the mechanisms—all this because we lack a biomarker that can tell us what's happening and who has this condition or not."

This lack of any standardized definition of what *it* is tends toward absurdity—even among scientists themselves, not to mention the public. Fasano's team at Harvard recently tried to get an estimate of how common gluten sensitivity is. The researchers simply defined it as someone feeling sick when eating gluten and feeling better when not eating gluten (and has tested negative for wheat allergy or celiac disease).

"We extrapolate an estimate of 6 percent," Fasano said, though as soon as the number leaves his mouth, he offers caveats. "Again, it's an estimate, it's very early."

Of course, with a definition so broad, this is essentially a survey of people's understandings of themselves.

"There's no evidence for that [rate] at all," countered Peter

Green, director of Columbia University's Celiac Disease Center. The rate of people who *think* something does not make it the rate of an actual condition, he reasons, veering into philosophy. To truly know what is causing this vast array of symptoms, Green and Fasano look forward to large studies in which everything in the subjects' diets and lives is controlled—a "double-blind" study. That could separate real effects from placebo ("which is *real*," Fasano emphasizes). There is also the inverse, "nocebo" effects, where people who believe they are doing something harmful, like eating gluten, will experience symptoms.

Also in 2016, another publicist invited me to speak with "Dr. Peter Osborne" about his "natural treatment of chronic degenerative diseases, with a primary focus on gluten sensitivity." The publicist touted his new book as "the first book to identify diet—specifically grain—as the leading cause of suffering."

In truth, the first books to claim that wheat is at the root of widespread suffering—not only the aberrant immune reactions in a small number of people who have wheat allergies or celiac disease, but everything from Alzheimer's to depression to heart disease—came out beginning in 2011 with the works of writers William Davis and later David Perlmutter. Like political demagogues, they sold antiestablishment views and revealed ostensible truths that no one else was willing to say—that the mainstream doctors *didn't want you to know*. Osborne makes similarly sweeping claims, writing that for people who are in pain, "removing all grains from your diet is more effective and safer than prescribing drugs."

Yet he has formed a group called the Gluten Free Society, which sells an array of supplements that, among other things, promise to "boost the immune system" and "detox." The largest words on many labels are "gluten free." Similarly, during the launch of Perlmutter's second bestselling diet book, he sold

"Empowering Brain Formula" pills on his website. I inquired about the practice with his book publisher (Little, Brown), who touted him as a leading expert on brain health and diet (both literally and by the act of publishing and publicizing his theories). The publisher did note in reply that "his online store is soon to be closed." William Davis continues to sell a "10-Day Grain Detox Course" for just $79.99. He launched a magazine called *Wheat Free Living*. His publisher, Rodale, keeps his diet books stocked internationally and boasts that the contents are "informed by cutting-edge science."

The willingness of people to buy these products and espouse these beliefs is most likely a symptom of real problems within the medical establishment. At a nutrition symposium at Tufts where I spoke in 2015, physician Douglas Seidner explained how vexing it is to be confronted with a stream of patients who are ignoring the most credible of dietary advice and yet religiously adhering to what is unsubstantiated. Seidner is director of the Center for Human Nutrition at Vanderbilt University, where he has been a practicing gastroenterologist for twenty-five years. His approach to gluten sensitivity relies on a study of the history of medicine.

Seidner explained that even as recently as when he was in training, a couple decades prior, a doctor's recommendations were taken as decrees. Medicine was governed by expert opinion and "evidence-based medicine," where doctors informed a patient on the best course of action based on biomedical data. But things have changed in the intervening years. As the ethical principle of patient autonomy has gained recognition, medicine has lurched toward patient-doctor collaboration. The movement is called shared decision making.

The concept of empowering patients is indisputably beneficial on paper, but it has proven challenging to execute. The shift from paternalism to stewardship is apparent nowhere more than with the people who come to Vanderbilt convinced that they have non-celiac gluten sensitivity. They request the blood tests that they've heard about from companies like Cyrex Laboratories or from the naturopaths such companies employ or from the demagogic booksellers who deal in fear and silver bullets. Seidner's patients come armed with their own opinions, charging doctors to present evidence if they disagree. If the doctor isn't persuasive enough, patients may go to another (possibly an "alternative") practitioner, who will tell patients what they want to hear, give them the label they seek, order the tests they want.

Two very small randomized controlled trials have so far studied what happens when people with vague gastrointestinal symptoms (commonly labeled irritable bowel syndrome) switch to a gluten-free diet. In the first study, thirty-four patients on gluten-free diets were told to add a muffin and two slices of bread into their normal routine for two weeks. Half of the people received gluten-free bread and muffins. By the end, that half reported fewer symptoms of pain and bloating than those who ate the standard bread and muffins. They showed no increase in antigliadin antibodies, however. The researchers concluded that "non-celiac gluten intolerance may exist," though they found no clue as to how it might occur. Two years later, Australian researchers repeated the study with a similar size and design and found no difference in symptoms between people who ate gluten-free products and those who didn't.

Fasano, Seidner, and Green do not recommend that people take themselves off gluten without first consulting an expert. "That would be something that would create a major problem," Green said, in that it may cause people to overlook the actual reason that they're sick. At Columbia, about half of people who come to see him believing they have a sensitivity to gluten end up having another condition. Generally speaking, he recommends against going gluten free unless you have celiac disease or wheat allergy, as the diets are usually less healthy overall. People who follow gluten-free diets often end up replacing whole grains with highly processed products in which sugar and sodium must be added to make up for the insipid textural experience. People also end up equating "gluten free" with "healthy." For the vast majority of people and products, the implication is rather the opposite. A gluten-free bagel from the brand Glutino, for example, has 43 percent more sodium, 50 percent less fiber, and 100 percent more sugar than a Thomas's Plain Bagel. It can get away with these levels because it lives in an aisle labeled "gluten free"—where that claim becomes all that matters.

The very existence of the words "gluten free" on a package or menu implies that gluten might be worth avoiding. These words, and their absence, can cause, alleviate, or exacerbate sensitivity. It's also possible that gluten-free endeavors leave people feeling truly better (or worse) because, in the process of avoiding glutinous foods, people make other changes to their diets.

"When people say they switched to gluten free and now they feel better," Fasano posited, "are we sure they just weren't eating in a broadly unhealthy way before, and now they're eating a healthier diet, and that's the reason they feel better?"

"I see a lot of people who put themselves on very restricted diets," said Green. "They avoid gluten, they avoid soy, they avoid corn—it's just unclear, to my mind, what that is about."

At least some, if not most, of the allure of this "elimination diet" approach is the sense of control it offers. Elimination diets appeal to our very normal desire for straightforward, attainable directives to protect ourselves from disease. Cut out this one thing, and you'll feel better.

"It's about much more than what you're not eating. It's about what you *are* eating," said Green.

The consensus recommendation to follow whole-plant-based diets is there, and it promises favorable odds. But the commercial interests behind it are too few. Without unduly fetishizing the past, something has changed that has brought about a rise in celiac disease.

Increasing concern about gluten sensitivity stands to overshadow the fact that celiac disease is quickly becoming more common, and no one knows why. It is around four times more common today than it was in the last half century—not just more commonly diagnosed, but in fact more pervasive, Fasano and Green agree. At Harvard and Columbia, both research centers have just pivoted toward a focus on the role of the microbes that live in and on us and constitute most of the DNA in our bodies. Massachusetts General Hospital recently launched a project to attempt to understand celiac disease and gluten sensitivity. Called the Celiac Disease Microbiome and Metabolomic Study, its goal is to stratify the population for targeted intervention—and to prevent celiac disease from developing altogether. The project will require hitting multiple moving targets at the intersection of the genome, microbiome, and life. We are far from understanding most any disease at that level.

"When I was a child, we used to eat seasonally," said Green. "Now we eat everything all year round." He mentioned the rise in C-sections, the inordinate use of antibiotics, and declining exposure to illnesses and natural environments as factors affecting our microbiomes and our well-being.

Fasano, for his part, cites breast-feeding, quality and quantity of gluten exposure, and the time at which gluten is introduced into a child's diet—all of which may have some merit in the rise of celiac disease, but none that is as yet a determinant by itself.

If I came away with a universal recommendation on what to avoid and be wary of, there would be two things, and neither would be gluten. The first would be hubris. As Green put it, "So many people think they know everything, and they just don't. We just don't." The other would be impressionability. Just because we don't know everything doesn't mean that we know nothing. Misunderstanding that distinction is, I think, at the heart of every diet trend.

"Celiac disease is the best-understood autoimmune disease, and there's so much we don't know," said Green. "It's just proving much more complex than I think anyone imagined. There are so many factors that we don't know, beyond genetics and food, our microbiomes, our immune systems. Everyone in every specialty is looking at the microbiome now. It seems to be involved in every disease. But we don't know how it's disrupted, much less how to undo that disruption."

In an early endeavor to contribute to that understanding, I told Fasano, I had gotten my microbiome sequenced. That's uncommon right now, but likely to become less so, as the cost of having it done has fallen from $100 million to $100 over the past decade. I had to mail a cotton swab laden with feces (my own) to the company, a tech start-up in San Francisco called uBiome, where they broke down the DNA therein and identified some of the bacteria that reside in my bowel. (Or that, you know, *are me*.) The microbiologist Rob Knight writes that "it might soon become a routine medical procedure ordered by your doctor." It could tell us all kinds of things about the gut-brain-microbiome-immune axis, though at the moment it is a novelty. The microbiologists at uBiome told me that based on my results—and what little is now known about gut bacteria and what certain ratios seem to indicate—they expected me to be obese and depressed, neither of which I am. I relayed that to Fasano, and he laughed.

"Well, that tells you how complex this situation is."

The microbiome results now are just statistical guesses, looking at correlations between bacterial populations and people with certain conditions. But they may very well be the key to understanding the remarkably complex interplay that leads to a vast spectrum of symptoms—those that many people today blame on gluten.

"The actual mechanisms are going to be much more nuanced and involve many factors," said Fasano. Like all of health and life, "if you dissect one piece, the risk is that you don't see the big picture."

Eggs versus oatmeal

In the middle of his career as a cardiologist at Harvard, Dariush Mozaffarian decided that rather than prescribing cholesterol-lowering drugs and opening clogged arteries with balloons, he could do more good by devoting his career to helping prevent those conditions. To him, the obvious way to do that was using food.

He's now the dean of the only sovereign school of nutrition in the country, the Tufts University Friedman School of Nutrition Science and Policy. If you're the sort to get nervous eating around health experts, dining with him is tough. We had breakfast once in Colorado, and he got an omelet, and so did I, but I asked for just the egg whites. He asked why I did that. My answer was that . . . I do illogical things all the time. I know that eating cholesterol doesn't seem to be a significant risk factor for any particular disease anymore. But the conventional wisdom is baked into my head.

I didn't get into my *other* egg thing. I had been in Boston for a

symposium at Tufts earlier that year. While I was there, there was a huge conference happening. Called Experimental Biology, it's *the* conference for people who study life. Every year it brings together thousands of scientists from across disciplines—anatomists, biochemists, molecular biologists, pathologists, nutritionists, pharmacologists—to present their new and sometimes exciting scientific research.

Because I like to pose hypothetical battles between things, a Tufts student called my attention to one of the presentations that might be of interest: eggs versus oatmeal. New research would be revealed that suggested a winner to the question so many have faced.

The presentation was definitively in favor of eggs, with the researchers reporting that they seem to *improve* people's cholesterol profiles. That was surprising. It's clear that the link between eating cholesterol and having high cholesterol levels in one's blood is much weaker than we used to think, and many experts (like Mozaffarian) believe it's nonexistent. But it was unexpected to find an inverse relationship.

So, what does it mean that eggs are healthier? I got in touch with the researchers, who were based at the University of Connecticut and led by Maria Luz Fernandez. She heads the Department of Nutrition Sciences' graduate education program and researches the effects of diet on heart disease. Since earning her PhD in 1988, Fernandez has served for several years as an editor of the *Journal of Nutrition* and half a decade on the Food Advisory Committee to the U.S. Food and Drug Administration.

In the eggs-versus-oatmeal study, she explained to me, her team gave people oatmeal for breakfast every day for four weeks. And then eggs for four weeks. At the end, 70 percent of people had higher levels of the "good cholesterol" HDL during the egg period. She specified that everyone ate *two eggs* per day for four weeks in a row. "People don't normally eat so many eggs," she said. "Especially women."

Fernandez has been studying eggs since 2002. She has even done YouTube videos explaining the benefits of eggs, I discovered, presented by someplace called the Egg Nutrition Center, in which she extols eggs because they contain the antioxidant lutein. It's an interesting argument, in that plants have many more antioxidants than do animal products. So I wasn't surprised by her answer when I asked if she eats eggs herself. She said yes, probably six or seven per week.

I dug deeper. What kind of oatmeal did they use? The Instant Quaker Oats packets (which are loaded with sugar)? Yes.

Did the study have a control group? No.

Our conversation left me only more confused. If she wanted an honest answer, why did she compare eggs to garbage-candy oatmeal?

After we spoke, I wrote to ask about any potential conflicts of interest she might want to disclose.

She replied, "Well, I have been funded by the Egg Board to carry out most of the studies."

So it's good that I asked. The fact that the research was presented at Experimental Biology and came from the University of Connecticut gave it credibility. Unconvincing as the egg-versus-sugar-packet study turned out to be, it illustrates the question that confronts so much of science today, especially in the realm of nutrition.

The American Egg Board is the trade organization for egg farms with more than seventy-five thousand hens. You know those twee commercials with the jingle "the incredible, edible egg"? (The latest iteration features Kevin Bacon and his brother being self-consciously twee.) That campaign comes to us from the American Egg Board. The Egg Nutrition Center, the source of those YouTube videos, is the "nutrition education division" of the American Egg Board. Its stated mission is to be "a credible source of nutrition and health science information . . . related to eggs."

One of the least credible things an industry can do is describe itself as credible. I'll admit I didn't vet everything in their section called "Egg Educational Materials and Science," but still, I couldn't find one word on their site that suggested that eggs were anything less than morsels bequeathed to us directly by a truly benevolent God. ("Eggs can play a role in weight management, muscle strength, healthy pregnancy, brain function, eye health and more.")

Even if every word were true, it would be an incomplete picture at best. It was major worldwide news in 2012, for example, when a large study from Western University of Canada compared egg consumption to smoking. Researchers looked at data from twelve hundred people and isolated the two behaviors. Smoking has been well proven to cause strokes and heart attacks by contributing to atherosclerosis (plaque buildup and hardening of the arteries). The Canadian researchers found that regularly eating eggs was about two-thirds as harmful, along those lines, as smoking.

Fernandez dismissed their findings out of hand: "People get passionate. But I believe eggs are healthy. That's why I'm trying to do my studies."

The question central to egg science, and to all science, is this: If the Egg Board funds research projects, does that necessarily mean the findings aren't credible? And who should we expect to fund egg research if not the people who have a financial interest in the outcomes?

The National Institutes of Health exist to eliminate that problem, doling out grants to researchers from a pot of taxpayer money. Their budget is modest and stagnant. Meanwhile, the number of scientists and questions and commercial products is only growing. So researchers wishing to study the health effects of eating eggs, like Fernandez, rely either on philanthropic organizations or on industries like the Egg Board. Throughout medicine, collaboration with the pharmaceutical and medical device companies is commonplace, often unavoidable. Conflicts of interest exist whenever enormous industries stand to gain by finding evidence in support of their product.

This doesn't mean that the Egg Board "bought" these findings. Fernandez and her researchers' study could well have come out against eggs. But none of Fernandez's work yet has. If it started to, what would happen to her funding? For an academic researcher, where career stability is tied strongly to the number of studies you publish, it's enticing to ally with an industry that will keep you prolific.

Only from an impoverished view of humanity could we assume that most scientists will lie and willfully misrepresent findings simply to favor whoever funds their research. It requires a much less impoverished view—a sympathetic view, even—to recognize the biases born of our instinct for self-preservation.

This turned out to be a question much bigger than eggs. An evangelist of the egg industry is not just leading the graduate education programs at a major university, but also editing an academic journal and consulting with the FDA. The question of eggs becomes a question of epistemology. Not, are they healthier than oatmeal, but instead: Who can I trust? How is knowledge acquired and propagated? Do people really eat those little oatmeal packets and think they're healthy?

I talked with the director of the National Institutes of Health, Francis Collins, about the problem of limited funds and increasing competition among scientists. "Some of that's kind of good, it inspires," he said. "But some of it's destructive. We're trying to get that balance right. Maybe it's been a little bit too much in this hypercompetitive space and we ought to relax a little bit and remember what we're trying to do here, which is to advance knowledge about life science and help people."

Collins's grand vision for biomedical science is a future of transparency and collaboration. One that rewards not just the number of studies that people manage to get published in scientific journals, but their collaborative spirit. In a more collaborative world, earth-shattering journal articles wouldn't be the currency of fame and fortune; the *entire process* of science—whether the findings are game-changing or banal—would be similarly regarded and rewarded.

"Maybe when you publish your CV, it doesn't just have the papers you published in *Nature* and *Cell* and *Science*," Collins suggested, "but it also has a reference to a database that you put up. Or a major data set that you contributed to a public database. And how many citations and how many downloads it has had. Because that's the way our field is going."

One area of research emphasizing collaboration is the U.S. government's Precision Medicine Initiative. It involves extensive cooperation between publicly funded scientists and private industry. Collins would like to see a "data commons" where people share not just their conclusions, but all the data they acquire. It would spare us from having to speculate about the conclusions of a study funded by the Egg Board, for instance. The organization and the scientists it funds could have a role in contributing data from egg consumers to a database, where scientists around the world could analyze and vet and compare and expand upon that data. If the Egg Board truly believes that eggs are as healthy as it claims, why not allow innumerable researchers around the world to have their way with the board's egg data? To analyze egg consumption alongside every other facet of people's diets, and how much they exercise and how well they sleep and exactly what their genome looks like. Then, maybe, we could begin to fully understand how to compare eggs and oatmeal.

In the interim, though, people must eat. We know enough that our decisions are not meaningless. Near the University of Connecticut, Fairfield University biologist Catherine Andersen, who studies how to use food to help prevent disease, captured the complexity of the egg/health relationship in a 2015 review of the best science on how eggs affect the body. She reminds us that eggs are about much more than cholesterol. And they affect every person differently.

She writes, "Bioactive egg components, including phospholipids, cholesterol, lutein, zeaxanthin, and proteins, possess a variety of pro- and/or anti-inflammatory properties, which may have important implications for the pathophysiology of numerous chronic diseases and immune responses to acute injury." Which translates to, essentially: Eggs are complex structures, and the physiological effects of digesting those structures cannot be distilled to a simple yes/no. Andersen reports having received no funding from the egg industry.

For those who prefer practical answers: Whole-grain oatmeal with nuts and fruit is recommendable to any (nonallergic) human, both from a nutritional and a societal perspective. Producing oats is easier on the soil, water, and air than is the factory farming of antibiotic-laden birds raised shoulder to shoulder in darkness who never learn to walk. Were we to raise these chickens without antibiotics, on a free range, and every one of the seven billion humans were to eat two eggs per day, the entire planet would be covered in chickens. Everywhere you look, there would be chickens. The oatmeal-versus-egg comparison is no more valid than an oatmeal-versus-diamond debate.

It's interesting to consider—eating diamonds. Someone somewhere will surely tell you it's beneficial. Before you go that route, see if that person owns a diamond mine.

Do probiotics work?

"Probiotic" products exploded as a trend after the Human Microbiome Project was completed in 2013. By 2020, the industry is projected to reach $46 billion. The concept is that you can buy and consume microbes that will favorably affect the ecosystems in your gut. As likely as it is that manipulating our microbiomes will revolutionize human health someday, it is not today accomplishable with probiotic products. That era of medicine is being presaged by years of entrepreneurial promises of what is not yet possible.

Today's products are sold under the supplement loophole, so there is no burden to prove that they do anything (or that they are safe). There is no burden even to prove that a product contains the bacteria it claims to contain. Microbes require very specific environments to survive, so, especially on a store shelf, they are unlikely to be alive when you buy them. If they are in there, there is no way to know how many of which microbes will affect whose gut microbes in what way. Consuming the "probiotic" products on the market right now is like reaching into a bag labeled "Assorted Seedlings" and taking a handful and throwing them into a forest. Some of them might not even be seedlings. If some of them do grow, will they be good for the forest?

Throwing seedlings into your own personal forest may not hurt. But there are some things that we know *are* good for our microbiomes, and they're the things that humans have long known are good for us. When experts use the newer term "prebiotic," they refer to substances that are prone to harboring a diverse, robust microbiome. The most effective known prebiotics are high-fiber fruits and vegetables. Peter Turnbaugh at Harvard has shown that diets high in meat and cheese rapidly and dramatically change microbiomes, limiting diversity and otherwise boding ill.

Discovering the extent of our microbiomes has yet to upend conventional wisdom about what's good for us, but it has begun to explain what we've long known to be good. The microbiologist Rob Knight, who led the Human Microbiome Project, likens "I'm taking a probiotic" to saying "I wasn't feeling well, so I took a drug. I heard that drugs can help." In his book *Follow Your Gut,* Knight concludes that the probiotic concept is promising for use in obesity and irritable bowel syndrome, but most anything being sold today under that label is based on "more hype than solid research." He offers glimmers of hope from early research: The probiotic *Lactobacillus helveticus* did seem to reduce anxiety in mice, and another quelled their "OCD behaviors." *L. paracasei* and *L. fermentum* appear, in early studies, to help some people who have atopic dermatitis. (I will have to tell Kaspar Mossman, who still itches.)

If vitamins offer a reflection of our psychology when it comes to health, then history may well repeat itself with probiotic supplements. Because some bacterial products seem to potentially be beneficial to some people, advertisers will lump all probiotics together, deem them *good,* and consumers will want them in all types in infinite quantities.

There will soon be a flood of products sold as probiotics that will ostensibly perform specific functions and address specific maladies. The people who buy these products will be seen not as fools but enlightened, on trend. Less trendy but cheaper are the methods clearly proven to maintain and diversify our microbiomes: avoiding unnecessary antibiotics, eating fiber, being outside, and chilling sufficiently.

How much worse is high-fructose corn syrup than "real" sugar?

High-fructose corn syrup is a study in linguistics and perception. Its greatest fault is that it bears a monstrosity of a name. One that conjures hyperprocessing and industrialized evils at their most extreme. But for all intents and purposes, it is the same as sugar that comes from sugarcane, which is equally hyperprocessed but still considered to be "real" sugar.

The corn industry knows this, and has been attempting to allow the name to be changed on food labels. A testament to the power in a name, cane sugar producers (American Sugar Refining) sought $1.1 *billion* in damages in 2015 from the Corn Refiners Association over a campaign in which the corn refiners called high-fructose corn syrup "corn sugar" and "natural."

The name *high-fructose corn syrup* is a historic artifact. The product does contain more fructose than older formulations of corn syrup—hence the "high." But high-fructose corn syrup actually contains a lower percentage of fructose than honey or agave syrup. All of these sweeteners contain some combination of fructose and glucose. A small number of scientists believe that fructose is worse for human health than glucose, most outspokenly Robert Lustig, the chair of pediatric endocrinology at UCSF. Conversely, a faction of scientists believe that prioritizing foods with a low "glycemic index" (containing more fructose than glucose) is preferable.

If we one day find ourselves so svelte that it becomes a worthy endeavor to parse the ideal ratios of fructose to glucose in our diets, then it's possible we accidentally traversed universes. Most nutrition experts do not see a practical point in making a distinction—sugar products are fundamentally the same in the effect they have on the body at the (minimal) quantities we should consume them. Foods that tout "no high-fructose corn syrup!" are dangerous in that they appear to be making a claim to purity or health, when all they are doing is demonizing one sugar as a tool to sell another. The effect is to distract from the clear knowledge that added sugar of any kind is a prominent factor in the leading cause of death worldwide (cardiovascular disease).

In the 2015 lawsuit over who gets to call their product "sugar," the corn refiners countersued for $530 million. The dispute was settled out of court. But the cane sugar producers have clearly won in the court of public opinion. The average American took in eighty-five pounds of corn sugar in 1999, compared with sixty-six pounds of cane sugar. Based on changes in public perceptions, by 2014 the score was sixty-one pounds of corn sugar to sixty-eight pounds of cane sugar.

Also capitalizing on the trend are the sellers of products that are sweetened with "juice concentrate"—sugar extracted from fruits—or else that deliver sugar in the form of honey or

TYPES OF SUGARS

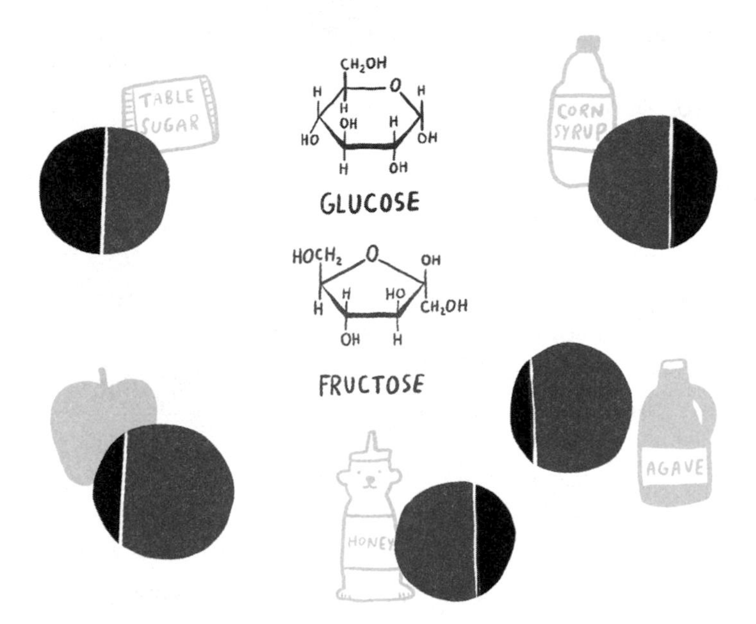

agave extract. It might be clearer if there were a convention in the naming: corn sugar, cane sugar, bee sugar, agave sugar, fruit sugar.

But even if the difference is academic, corn syrup *is* an important concept in terms of the role of the technology behind it, which made cheap sugar ubiquitous. The United States created an agricultural system that produces and subsidizes tremendous surpluses of corn. The ability to swiftly turn that corn into sugar led to more sugar in the country's diet, at lower prices than cane sugar. There are plenty of reasons to rage against corn industry subsidies—essentially the use of tax dollars to keep cheap sugar in our diets—but there are also reasons to rage against the cane sugar industry, which is more immediately threatening to the environment than the production of corn sugar. How we vote with our money in this contest is a meaningful decision, but it's not really about our individual health.

What if my tongue ring came out and I accidentally swallowed it?

It would probably be fine, but any time you swallow something sharp, the concern to doctors is that it could perforate the wall of the bowel. If it did, the same bile and intestinal flora that so diligently keep us healthy and sane would spill into the abdominal cavity and cause a life-threatening infection. Maybe that's why tongue rings are so cool, because they mean you're living on the edge. A middle finger to the notion of a perforated colon.

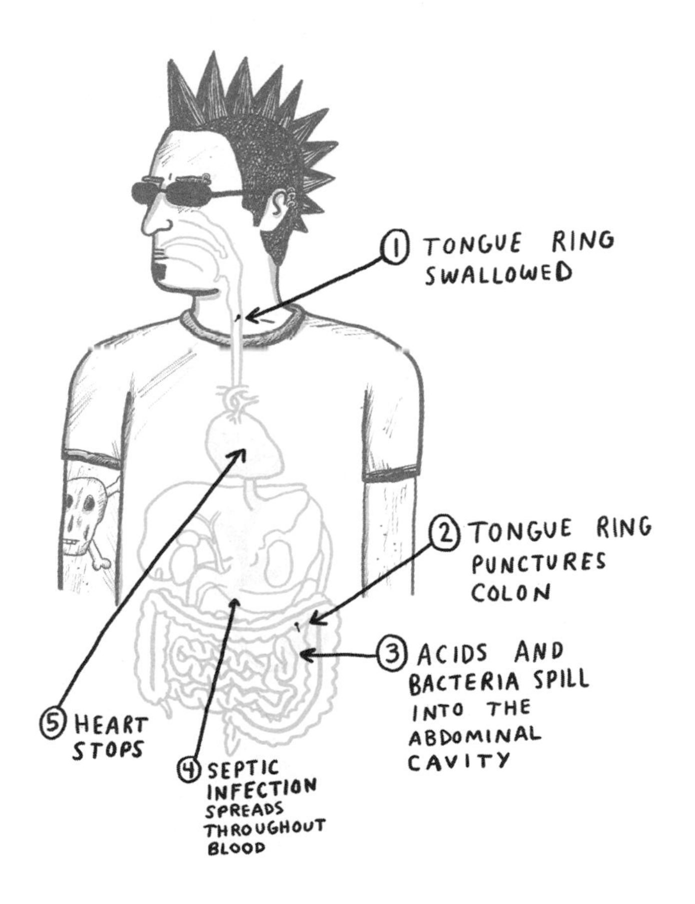

I need dairy or else my bones will break?

The most indelible answer I've heard to the dairy question came from an interview I did with Russell Simmons, the co-founder of Def Jam records. Once addicted to heroin and angel dust, the bald yogi is now an advocate for health-promoting lifestyles. He meditates every day and posts pseudospiritual, sometimes painful aphorisms on Instagram. ("You get there by realizing you are already there.") He is also a sworn vegan.

Simmons's reasoning is that even cows don't drink cow's milk. It's true that if you sit and watch a field full of cows for a prolonged period of time, as I have, none of the adult cows will start suckling at another's udder. Calves drink milk until they're old enough to eat solid food, and then they stop. They do not produce lactase, so adult cows are, like most mammals, not tolerant of the sugar lactose. Humans might seem to operate the same way with milk from our own species—it's beneficial at one point in life, but less so at another. It's enough to get you in trouble with the good people in human resources if you tell colleagues that you're drinking human milk and that you find it delicious and necessary to keep your bones from fracturing.

So why is cow's milk so entrenched in Western cultures—not as a delicacy appreciated by some, but as a daily staple that many believe is integral to proper body maintenance?

This is a question not really about calcium and osteoclasts as money and politics—a question that, to think about at length, can itself fracture a person's bones. The case for milk is a reductionist one: It always comes back to the suggested benefit of calcium, phosphate, and vitamin D, typified in a 2013 research review that concluded, "Dairy foods are important sources of these nutrients."

This is careful wording. Once you start noticing claims that some product "is an important source" of something, you see they're everywhere.

In the 1920s it was understood that some children would grow up with bowed legs. Rickets had been a common part of life since industrialization brought people to cities, which led them to get less and less sunlight. Eventually, British scientists noticed that rickets could be prevented by consuming yeast that had been exposed to ultraviolet light. The light converted a steroid, ergosterol, into a compound that would come to be known as vitamin D—the deficiency of which caused warping of the bones. So Britain began adding ergosterol to milk and irradiating it. Once they figured out how to convert ergosterol to vitamin D in labs, it was added directly. Within years, rickets was essentially eliminated. It was a momentous accomplishment for public health, and one that would only solidify the perception that vitamins were some kind of magic. Food producers began adding vitamin D to any possible medium, including hot dogs and beer. ("It gives you the cooling tang that soothes heat-frayed nerves and awakens jaded spirits. Schlitz: The beer with Sunshine Vitamin D.")

Vitamin D ultimately causes our bowels and kidneys to absorb calcium. When our blood absorbs too much calcium, that can upset the heart's electrical currents and cause arrhythmias. Over time, calcium accumulates in the walls of our blood vessels, too, making them rigid pipes amenable to blockage (as in a heart attack). The kidneys will attempt to excrete the excess calcium, but some of it will get left behind and accumulate into kidney stones. Those we must pee out, which some people say is comparable in sensation to childbirth.

Many countries stopped adding vitamin D to milk in the 1950s, when children in Great Britain were found to have too much calcium in their blood. The country banned vitamin D fortification for most foods, and other European countries followed, keeping levels from getting out of control by allowing it only in certain staples, like margarine and cereal. Many European countries still do not add vitamin D to their milk.

Our bones do become weaker—less dense with minerals—as we age. As people today live to unprecedented ages, we do so on bones that are ever more frail and prone to breaking. And it is true that eating foods that contain calcium helps in keeping our old bones mineralized. Phosphorus aids in retaining that calcium. As does having enough vitamin D, most of which is produced in our skin in response to sunlight exposure.

But cow's milk is an example of how a system can arise around a belief (in this case, that milk strengthens bones), and how science can serve to perpetuate that belief. Within the current global food system, many people do rely on dairy products to get their calcium and phosphate (and in some countries, their vitamin D). Dairy is, in that sense, "important."

But it is only important within the food system we have created. Calcium, phosphorus, and vitamin D are all easily obtainable in ways other than dairy. Vitamin D is in many fortified foods even today. Many others contain as much calcium and phosphorus as cow's milk (calcium: spinach, broccoli, sesame seeds, actual oatmeal; phosphorus: beans, artichokes, lentils, avocados). It is odd to insist on cow's milk as the superior method of delivery, when countries with the highest milk consumption also have the highest rates of osteoporosis (United States, United Kingdom, Finland, Denmark). This is just a correlation, but not one without implications for the role milk plays (or doesn't play) in protecting our bones.

In my experience, the people who recommend that humans *should* drink cow's milk, specifically, are almost invariably those with ties to the dairy industry (which made $36 billion in revenue in the United States alone in 2015). Robert Heaney, a professor emeritus at Creighton University in Nebraska, spent most of his career advocating cow's milk, which he depicted as essentially the only plausible approach to bone health. In popular media, Heaney would drop dairy plugs at times that struck me as odd. In 2015 news coverage in *Time*, for example, Heaney said of seltzer, "It's not harmful. But if it displaces a beneficial beverage, such as dairy milk, then that's not good."

(All health recommendations might be made in a similar way, then, relative to their effect on your intake of dairy milk. Vigorous exercise is good for the heart, so long as it does not lead you to decrease your consumption of dairy milk.)

On the Creighton University website, Heaney made the interesting claim that lactose tolerance can be achieved with proper fortitude: "Someone who complains of severe lactose intolerance symptoms can almost always be brought up to the point of consuming three full glasses of milk per day without symptoms if they build up the exposure gradually." And so "don't let lactose intolerance get between you and your getting enough of the nutrients you need for bone health."

Heaney was a consultant to the dairy industry, which also funded his research. The home page of Creighton University's Osteoporosis Research Center featured an older woman smiling, with the breeze blowing back her blond-gray hair as she drinks a large glass of milk. The scientific review article noted above, which does not appear to be directly funded, cites seven of Heaney's studies.

At the Boston consensus conference of nutrition researchers, the only one who took the podium in defense of cow's milk was Steven Abrams, chair of the department of pediatrics at the University of Texas at Austin. His work has also been supported financially by the dairy industry, by which he has been employed as a consultant (currently with the Milk Processor Education Program, or MilkPEP, which is funded by milk companies and "committed to increasing fluid milk consumption"). Even still, Abrams speaks cautiously. The furthest that he went at the conference was to say that drinking milk is "a part of a healthy diet for those who choose to consume dairy," and positing the important hypothetical, "What would happen if we took milk and dairy products out of the diet? Calcium intakes in children would markedly drop. Vitamin D intakes would drop. Potassium intakes would drop."

That's true, assuming that everyone traded their milk for soda and continued to forgo fruits and vegetables. The speculation is critical, though, because Abrams was also part of the formal consensus committee that led to legislation by the government to continue subsidizing the dairy industry. Every five years, the U.S. Department of Health and Human Services convenes a panel of experts to review all nutrition research. The 2015 panel of fourteen academic nutrition scientists pored over all relevant research and produced a 571-page report about the optimal health-promoting diet. It recommended approaches to feeding people to optimize their health. The weight of evidence came down in favor, again, of whole fruits, vegetables, whole grains, nuts, and legumes, with scant mention of dairy as anything more than a possible source of nutrients.

But these recommendations are just that—recommendations. The federal Dietary Guidelines for Americans are ultimately issued by the U.S. Department of Agriculture. Tradition suggests that the department is to take the experts' report into account. But ultimately the principal interest of the Department of Agriculture is the prosperity of the country's agricultural economy. It is no small conflict of interest that this entity should advise Americans on what is best to eat.

This is a, if not *the,* fundamental conflict of interest in public health.

The Department of Agriculture did not accept many of the new recommendations. They went still further. During the hearing over the guidelines in 2015, congresspeople from the agriculture subcommittee outright chastised the secretary of health and human services, Sylvia Matthews Burwell, for choosing a team of scientists that they deemed to have an agenda against milk. (If anything, the committee had the opposite conflict of interest, with Abrams having received support from the dairy industry.)

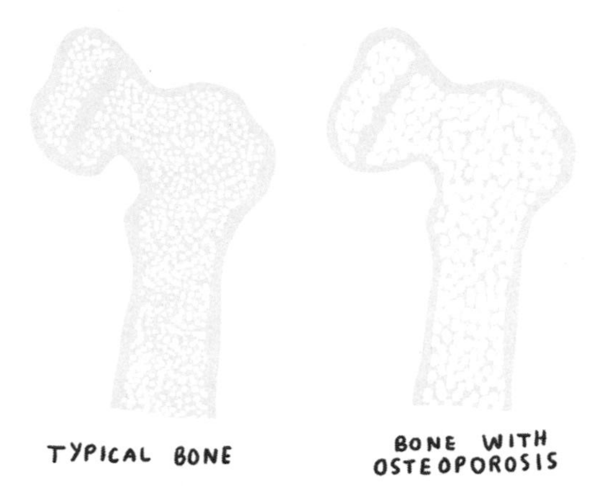

These outspoken congresspeople came, predictably, from states where the dairy industry was an important part of the economy (and a powerful lobbying interest). So dairy consumption remained a key element of the 2015–2020 Dietary Guidelines for Americans. They recommend that all adults consume three cups of dairy every single day.

These guidelines are much more than rules of thumb. They determine how tax money will be devoted to promote nutrition campaigns. (Like "Got Milk?," which is funded by MilkPEP and the federal government. Did you know you were paying to see those ads?) Even more important, the nutrition guidelines direct the billions of dollars that go to the poorest Americans, through programs like the Special Supplemental Nutrition Program for Women, Infants, and Children (WIC), and to fund school lunch programs. This is why cow's milk is prominently featured in every cafeteria in the country.

Frank Hu, a professor of nutrition and epidemiology at the Harvard School of Public Health who also served on the advisory committee, wears his exasperation on both sleeves. He was asked to lead the committee in its rigorous review, only to be accused of bias by the Department of Agriculture. It's especially frustrating for Hu, whose own research supports the recommendation away from milk (and who makes a priority of remaining free of conflicts of interest). When he compared the health effects of dairy fats to those of fats from other animal products, he found they weren't much different. When people replaced dairy fats with vegetable fats, though, they saw a significant reduction in cardiovascular disease. (Which is, again, the leading cause of death.)

Yet, ostensibly in the interest of our health, the federal government gives enormous support to the dairy industry and almost none to the production of fruits and vegetables. Which is not to say it's wrong to base your diet on what is best for the nation's existing politico-agricultural infrastructure. But it's not a diet fad that I hear a lot of people adopting consciously.

Are we made to eat meat?

We weren't made to do anything. Our bodies are collections of processes that exist as responses to other processes.

. . . I know, I'm just talking like a normal person. Should I be Paleo?

I know what you mean. I'm a normal person.

A philosopher might start with the observation that we aren't supposed to eat rocks. If we try to eat rocks, they chip our teeth and go undigested by our acids and enzymes, which are futile against them. They pass right through us. Hearing them clank against the porcelain, the person in the bathroom stall next to you might ask if you're okay.

So clearly we aren't "made to" eat rocks. By extension, then, there are things that the human body is not made to eat, and things that it is. A less extreme example: Clearly most of us did not evolve to tolerate the dairy sugar lactose. The burgeoning field of "ancestral health" is predicated on the idea that we

WHAT CAUSES OBESITY?

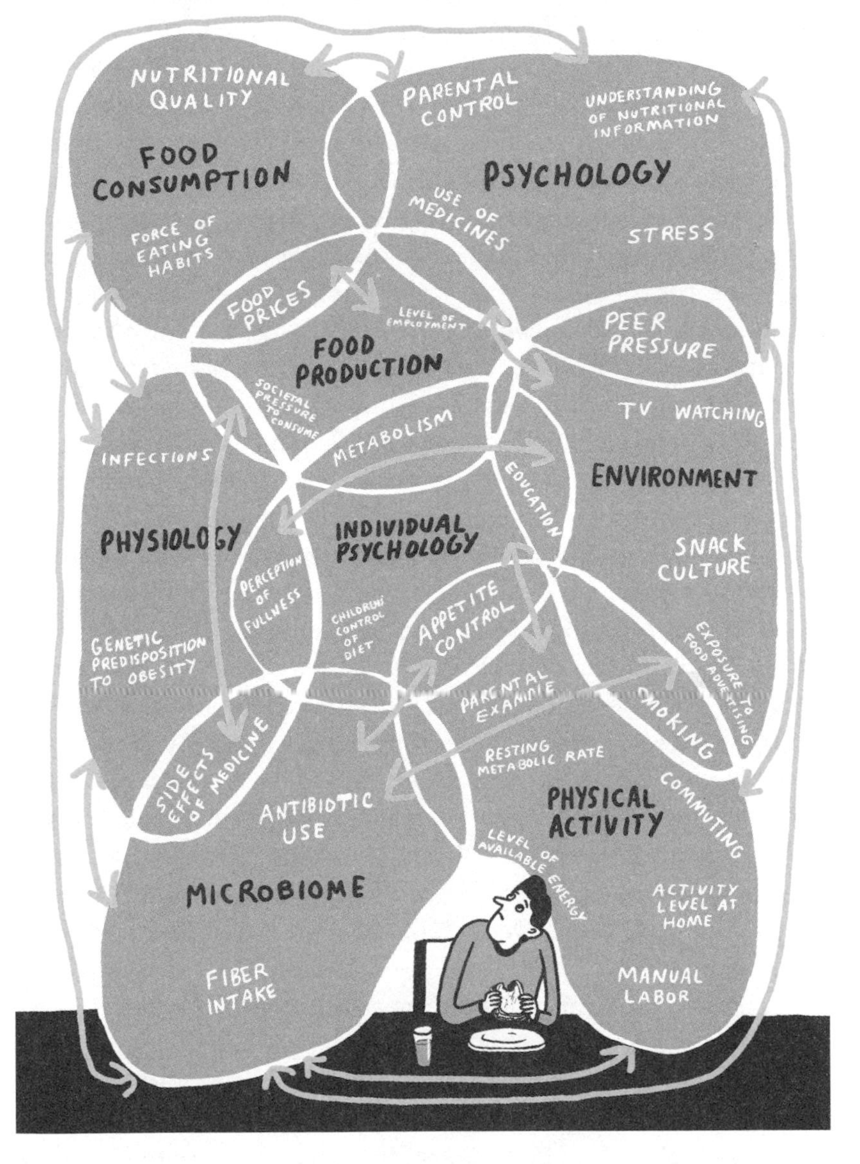

would all do well to learn from ten thousand years of human evolution about how our bodies are best suited to be used. If we do not have the bodies to digest certain things well, then maybe it's best to eat other things.

The meat question is less clear than the rock or the lactose question, because few of us experience immediate symptoms from it, such as the retching of a non-lactose-tolerant person who has just been milk-pranked.

The critical distinction is that tolerating a food is different from thriving from it. And unlike rocks, or poison mushrooms, say, there are things we tolerate in the short term that injure us in the long term. Like, we could seem "made to" eat cupcakes, in that they don't make us feel acutely ill. The opposite, actually; they make our brains release pleasure-generating dopamine. But we also know that over a longer term, cupcakes can destroy the ability of our pancreas to manage blood sugar levels, leading to heart attacks and strokes.

Inferring what's good for our bodies based on our anatomy is not a new idea. Two thousand years ago, the Greek historian Plutarch observed that "to none of the animals designed for living on flesh has the human body any resemblance." (Now who talks not like a normal person?) Plutarch reasoned that we have "no curved beak, no sharp talons and claws, no pointed teeth, no intense power of stomach or heat of blood which might help [us] to masticate and digest the gross and tough flesh-substance. On the contrary, by the smoothness of [our] teeth, the small capacity of [our] mouth, the softness of [our] tongue, and the sluggishness of [our] digestive apparatus, Nature sternly forbids [us] to feed on flesh."

This was three centuries before the first recorded human dissections, during which Greek physicians saw that our intestines are twelve times as long as our bodies. Lengthy intestines are a feature herbivores need to digest fibrous plants, while carnivores like wolves and bears have intestines that are many times shorter than ours, just about three times as long as their bodies. They have stomach acid that is many times stronger than ours, and stronger jaws. Our broad fiber-grinding molars were obvious to Plutarch, but only later did we learn that our saliva contains enzymes that begin to digest plant matter before it even enters our stomachs. The microbes in our guts (essentially an extension of what we think of as our natural anatomy) appear to thrive on a high-fiber (plant) diet, while low-fiber diets lead quickly to widespread microbial extinctions and loss of diversity.

None of this *forbids* us to eat meat, but it does mean that we look more like herbivores than carnivores. We do have isolated teeth on either side that come to something of a point (as do those of many herbivores). This confuses the proposition a bit, to some. The reformed cattle-rancher-turned-sustainable-farming-advocate Howard Lyman writes that when people argue for meat by pointing to their "canine" teeth, he invites them to try using them to "tear into the living flesh of a moose. I have challenged many people to do so, and not one has come back with the moose in his mouth."

Of course there are clear limits to that line of thought, namely, the rejection of all modernity. Most of us are nearsighted, for example, which doesn't mean we're "meant to" *not* wear glasses. During a recent visit to his lab at Columbia University, microbiologist Ian Lipkin warned me that this is the type of reasoning that leads people to reject vaccines. It is a seductive thing, even to intelligent people like Bob De Niro ("You know Robert De Niro?" I asked. "Oh, Bob?"), who advocated for a 2016 film that warned people against one of the most clearly and profoundly beneficial measures in the history of public health. Taken to another extreme, it would be difficult to argue that we were *meant to* use fire escapes or smartphones or X-rays.

That last one is known well to Stanley Boyd Eaton, a soft-spoken graduate of Harvard Medical School's class of 1964, who spent years scrutinizing X-rays (and, later, CTs and MRIs) as a professor of radiology at Emory University. Over the course of observing so much disease, he found his passion: studying the role of food in maintaining health. We discussed our similar arcs at the consensus conference in Boston. He believes that nutrition research is so difficult to do, and so little is yet known, that the field is essentially what the philosopher Thomas Kuhn would have called "a pre-paradigmatic, emerging discipline." It does not operate under a unifying body of principles that explains the distribution of disease among people. What it does have is an understanding of gross anatomy and cultural history. Eaton quotes Theodosius Dobzhansky's famous line, "Nothing in biology makes sense except in light of evolution." So it is largely from history that Eaton has concluded how people should eat, and that conclusion is that we should eat meat.

In 1985 Eaton and colleague Melvin Konner published an article in the *New England Journal of Medicine* that would become the foundation of a prevailing ignorance on a scale they could not have imagined. It was titled "Paleolithic Nutrition—A Consideration of Its Nature and Implications." This became the foundational document of what's known now as the Paleo diet. (Though it would be Eaton's colleague Loren Cordain who would go on to write seven pop diet books on the subject, approaching it from more toward what Eaton called the "entrepreneurial end of the spectrum.")

Paleo eating is predicated on what Eaton proposes to be the missing paradigm for nutrition science: to attempt to prevent modern disease by eating based on how our bodies evolved. To Eaton and his spawn, that means drawing from what the species before us ate during the Paleolithic era, beginning 2.6 million years ago, and mostly before the first *Homo sapiens* 100,000 years ago. The primary implication, as Eaton now explains it, is to avoid the refined grains and added sugars that make up so much of the modern human diet, because they are "totally alien to our biology." (He explains that *Homo* have long eaten honey, but they ate it on the comb, so it had fiber, like eating a whole fruit. Our blunt teeth can grind through it, apparently.)

Relative to the plethora of food products sold today that are primarily starch and sugar, he says, meat is less alien to the human gut. Of course, that isn't to suggest that biological harmony is implied simply by the fact that something is less alien to our bodies than, say, Oreo cereal. In practice, too, it is extremely difficult to eat like a Paleolithic hominid in a food environment where one cannot readily order mammoth. Thanks to centuries of breeding and an ever-changing global ecosystem, most plants and animals that go into our food today are as different from their forebears as we humans are from *Homo erectus,* which only figured out fire around halfway through its era. Our bodies have also changed genetically and epigenetically, as have our microbiomes. Still, the notion that there's wisdom in eating like our Paleo ancestors has been widely interpreted to mean that people should eat as many of today's enormous chickens and domesticated cows as we like.

What's most remarkable about this interpretation is that it is predicated on an understanding of an enormous swath of history, going all the way back to the Paleolithic era, but it *only* looks backwards. It does not consider the future.

So Eaton, the accidental originator of the movement, has come to see and speak out against it as an egregious error. He estimates that in the Paleolithic era, people ate about three times as many fruits and vegetables as modern humans do. This is widely overlooked by people who hear Eaton and colleagues say that we evolved to eat meat and then run immediately to eat as many animals as they can. During the Paleolithic era, there were scattered groups of *Homo sapiens*. There are now nearly eight billion. Eaton said clearly that having eight billion people eat a meat-centric diet is impossible (based on the land and water required to produce so much meat, and the environmental impact of the process).

In Boston, he engaged directly with the vegans—the famous advocates Dean Ornish and T. Colin Campbell, among others—about what he considers "the crux" of the Boston meeting: the simple fact that we are going to need "70 percent more food by 2050." (This number is thrown around all the time, but I've also spoken with environmental scientists who note that according to projections from the United Nations, the world population will grow to 9.6 billion by 2050. This is a 35 percent increase in humans. In a thirty-five-year period. Which is insane, and totally unsustainable, but for the record, it's not 70 percent.)

What happens to weight when it's "lost"?

It's converted mostly to carbon dioxide and breathed out.

"I think there is a Paleo diet that could be vegan, or close to it," Eaton pronounced from the stage, an intellectual olive branch. By the end of the day, the term "Paleo-vegan" was coming up a lot. Our protein will need to come primarily from plants or some synthetic source. Whether meat-heavy diets cause disease (which Ornish and Campbell clearly believe they do) is moot if the population grows to that size. We might as well be debating the health benefits of eating diamonds.

Even though eating like a mid-Paleolithic hominid is impossible, the general idea of using functional biology to inform modern health is intriguing. At the University of Oklahoma, anthropologist Christina Warinner studies the feces of ancient humans. She directs the Laboratories of Molecular Anthropology and Microbiome Research, which have the largest collection of ancient human DNA in the world. By analyzing the twenty thousand years of genetic information in these people's dental plaques (essentially fossils that form while we are still alive) and petrified stool clumps, called coprolites, Warinner can see what the microbiomes of ancient people looked like, as well as what organic matter these people ate. "Their foods were fairly fibrous, so you can actually see seeds and whole plant pieces in [their coprolites]," she explained to me, "and then, when we get the genetic data, that allows us to identify species."

"The 'Paleo diet'—as it appears on TV and in the news and books—really has no basis in the archaeological record," she explains. At the same time, the connections she has made do point clearly to our modern diets as the source of our disease. It's true that ancient people rarely lived beyond their forties or fifties, and that they were not paragons of health. But they died with little or no evidence of cardiovascular disease. They died in many cases of infectious diseases and accidents of the sort that are largely eradicated today. "We think there's an important connection," Warinner explained, "that there is something about this urban-industrialized lifestyle which is removing the diversity in our guts and as a result is making us more vulnerable and susceptible to these metabolic diseases."

In earning her PhD in anthropology at Harvard in 2010, Warinner didn't envision herself becoming a dietary expert. But as people have lately taken to looking to history for answers to the diseases of modernity—and rapid advances in DNA sequencing technology have made it possible to study the history of human health in unprecedented detail—anthropologists like her have come to play an important role in medical science. They are, essentially, gathering data from a global experiment that lasts not weeks or months, but thousands of years.

By looking at human variation on an evolutionary time scale, Warriner can determine when major diseases arose, and what specific human behaviors are contributing to them. We have proven to be an enormously adaptable and resilient species, capable of surviving in deserts and in the Arctic. Fifty thousand years ago, we weren't the only human species on the planet. But by the end of the last ice age, we were. Why are we the only ones who survived? Why did Neanderthals die out, and Denisovans, and *Homo erectus*—but we humans survived?

"I think that the answer is probably that we have an incredible dietary flexibility," Warriner told me. But that dietary flexibility goes only so far. "At what point does our diet change so much that our bodies can't keep up? And I think that that happens with industrially processed foods. I think that is kind of our breaking point—and you see that in our health."

With so much industrialization, fiber often fell out of the mix. "The whole point of dietary fiber is for our microbes," she said. Reducing fiber intake leads to a less diverse microbial community, which is "associated with a whole host of consequences for our overall health."

Meat contains no fiber, so meat could be at best a small part of an advantageous diet by that logic. But debating the exact percentages or type of meat is moot if its production only charts us on a path to extinction. Which it seems to be doing, even according to Eaton. He endorses the common estimates that the world population will reach ten to eleven billion by 2100. "They'll be 'existing,'" he says of the future generations, while "Earth's other life-forms will be going down the drain. However, if we simultaneously take measures to mitigate, to address the causes, we could decrease our population."

This is where the pop-Paleo movement becomes dystopian.

"It is entirely possible that by 2200, we could have one hundred million humans flourishing," Eaton continued from the lectern. "Not just existing, but flourishing. And Earth's other life-forms will be on the way back."

If convincing people to cut back on meat is difficult, try selling them on population control. He closed his lecture to the small group of nutrition scientists this way: "We have now a nonsustainable situation. I recommend that we adapt in the near future, as a partway measure, and that we mitigate in hopes that the far future will be optimal."

His preferred approach to mitigation is a *voluntary* one-child maximum. The vegan Ornish raised his hand and asked if Eaton was saying we could eat meat if we decrease the population by orders of magnitude. "That's hardly a compelling argument for eating meat," Ornish chided.

"Maybe not for you," Eaton replied. On the screen behind him loomed an image of a filet mignon.

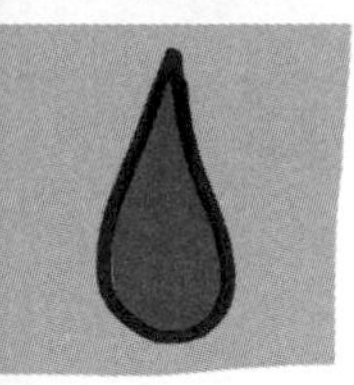

PART FOUR

DRINKING

THE HYDRATING PARTS

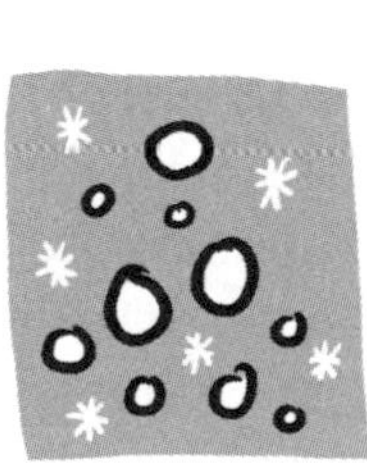

The entrepreneur E. J. Young was drilling for oil in Ohio in 1897 when he hit something else: the remains of an ancient lake that had dried up millions of years earlier. What was left was not oil, but a mineral, salt. In this case, it was an enormous table of salt, larger than any known in the United States. So Young did the sensible thing and went into the salt business.

The town above the salt, Rittman, was like many others in the Midwest in that it was born when the railroad came through (and named for a railroad executive). So all that Young had to do was create a company to manufacture boxes for the salt, and the railroad provided him easy means by which to become a powerful salt baron. In 1948, Young's salt operation became the one that most people recognize by name today: Morton Salt.

The label adopted a little girl under an umbrella and the claim that it "never cakes or hardens in any kind of weather." Because summer humidity was causing table salt to clump, Morton's set itself apart by adding magnesium carbonate, which kept it "pouring" even in July.

The chemistry worked, and the image persists, and Morton Salt is still largely drawn from the table below Rittman, home of the world's largest evaporating salt plant. The product graces the tables and kitchens of millions of people around the world. But even in his most entrepreneurial moment, E. J. Young did not likely predict that millions of people would have his sodium injected into their veins to keep them alive.

Among the most important medical interventions that exist today—and the most popular prescriptions in the world—is salt water. Put nine grams of salt into a sterile liter of water, and this is the multibillion-dollar product known as saline. It's injected into most patients who are admitted to most hospitals, and Morton is a primary supplier of the salt. What does not come through our mouths goes directly into our bloodstreams. There is truth to the company's marketing claim that "Morton is an American way of life."

One of the major producers, the pharmaceutical company Baxter International, ships more than a million units of saline alone every day. Pallets of saline are among the first things shipped to disaster zones. The global demand is so great that for much of the last two years the U.S. Food and Drug Administration has declared a shortage of saline.

Though a saline bag typically costs around one dollar, the shortage has lately called attention to practices at hospitals and clinics sometimes charging two hundred times as much. Include the charges for administering the fluid, and its cost can balloon to one thousand times the usual cost of a bag. In 2013, a hospital in White Plains, New York, charged one food-poisoned patient $6,844 for her minimal stay, *The New York Times* reported, including $546 for six bags of saline. The cost to the hospital would have been about $5.16.

Amid the ongoing shortage of these expensive bags of salt water, the American Society of Health-System Pharmacists made the novel recommendation that, whenever possible, medical facilities use "oral hydration." That is, of course, medical jargon for "drinking."

While most doctors are intensely trained in the art of hydrating people via the vein, the question of how best to drink is somehow open to interpretation by every company and celebrity who feels qualified to dispense advice—be it Taylor Swift grinning with a swath of dairy smeared across her upper lip, Rihanna pulling from a box of coconut water, or Michael Jordan grinning into a camera over an open bottle of lemon-lime Gatorade because "Nothing beats Gatorade." (Even though he also hit a nearly identical pose in another ad holding a can of Coke next to the slogan "Can't beat the real thing.")

Inside hospitals, hydration is a serious priority, approached meticulously. Doctors calibrate sodium concentrations and calculate precise rates at which the bag drips into our veins. Outside of hospitals, there is chaos. We hydrate based on how beverage companies see fit to implicate our kidneys in our quests to purchase purity and happiness. How, then, should one drink?

Do I need eight glasses of water a day?

In a blip in a commendable public health career, the First Lady of the United States, Michelle Obama, championed a campaign in 2013 called Drink Up. The premise was straightforward: "Drink more water." This was ostensibly because, as the nutritional adviser to the White House noted during the launch of Drink Up, "40 percent of Americans drink less than half of the recommended amount of water daily."

That calculation is a difficult one, because there is no recommended amount of water. Since I wrote about Drink Up for *The Atlantic,* I've pressed hydration researchers for an answer, and none will even attempt to give me a water quota. What they all say is that we drink too much soda and juice, and that's a primary cause of obesity and the diseases thereof. But the Drink Up campaign was co-sponsored by PepsiCo, which sells not only water and "enhanced water," but also soda.

Which may or may not be the reason that Obama and Kass artfully deflected questions on the subject of soda—"We're being completely positive. Only encouraging people to drink water, not being negative about other drinks."

With so much money and celebrity going into selling soda and juice, people of cultural influence do need to be negative counterpoints. While saying only that *more water is better* isn't likely to cause health problems directly, it causes them by omission. It also perpetuates a long-held misunderstanding about how hydration works. It's a misconception that I've found annoys hydration experts, when it doesn't infuriate them.

"I don't like eight glasses a day," Susan Yeargin, a physiologist at the University of South Carolina's athletic training education program, where she studies hydration and heat-related illnesses, told me. "It's not a good rule of thumb." In one video tutorial for athletic trainers, the young scientist explains how to remove heat from a hyperthermic person's body using an ice bath. ("The goal is to get the water as cold as possible. If you're starting with warm water, then you want to just keep adding ice.")

Better than an ice bath is preventing hyperthermia, or heatstroke, in the first place. Integral to that is proper fluid intake. Sweat cools the body by speeding conduction of heat away from the skin. When the body is dehydrated, it sweats less and overheats quickly.

Yeargin advises her athletes that keeping track of their urine color can help gauge their hydration status: "Light yellow means you are hydrated, bright yellow means you're dehydrated, and dark yellow like apple juice means you're really dehydrated. People should drink fluids in order to stay light yellow." Our pair of bean-shaped kidneys keeps the body's electrolyte levels constant by excreting water, electrolytes, and nitrogen. The varying concentrations are reflected in urine color, and indeed urine color (and taste) has for centuries served as a proxy for human health.

H_2O is a chemical added to our bodily equations just like anything else. There is no reason to hold it apart, as many of us have been taught to do, and think that more is always better. Even water can be deadly in the right doses.

If you've ever tasted blood, which is okay to admit, then you know it's salty. The concentration of that salt (much of which comes from Ohio) is among the most important numbers in our bodies. Hydration is not a story of giving the body water; it is a story of giving the body the materials it requires to keep itself in balance.

How to embrace sweating

Sweating is a perfect example of the way we tend to be self-conscious about the most important and ingenious things our bodies do. Even though sweating can lead to dehydration that can be fatal, death would have come considerably more quickly if your body overheated because you couldn't sweat. We sweat because wet skin cools faster than dry skin. The liquid on the skin's surface helps conduct heat away from the body. Going outside in fifty-degree air is no big deal, but jumping into a fifty-degree lake is rough. It's all because water conducts heat away from us faster than does air. Sweat on the skin facilitates cooling. So it's better not to wipe sweat away. Just let it linger and accumulate. It's a beautiful work of physics that lets the body achieve balance despite what we put it through.

For all that we eat and drink and sweat and urinate, our kidneys do a phenomenal job of that. Eat salt, and your body retains water to keep the sodium concentration in your blood from getting too high. (So we feel thirsty.) Kidneys can almost always manage to keep our blood sodium levels at 140 millimoles of sodium per liter of plasma. Keeping that concentration from getting too high or too low is the core tenet of hydration, and thus life.

When we sweat out a lot of salt or drink too much water, sodium levels in our blood begin to fall. This can lead to the dangerous condition called hyponatremia, heralded by lethargy, restlessness, confusion, drowsiness—even seizures and death in severe cases.

In wealthy countries, the occasional extreme case of water intoxication makes news—like marathon runners who flirt with death, or fraternity pledges whose brains terminally swell while doing the "Gallon Challenge" (chugging a gallon of water as quickly as possible). The most common cause of symptom-inducing water intoxication, however, is a phenomenon that doctors call *psychogenic polydipsia* (*psychogenic* = originating in the mind, *polydipsia* = much drinking). This "compulsive" water drinking is reported in between 6 percent and 20 percent of psychiatric patients. It's especially common in people diagnosed with anorexia nervosa, psychotic depression, and bipolar psychosis. The impulse is sometimes a symptom of one of these psychiatric disorders.

But sometimes the water intoxication is itself responsible for the psychiatric symptoms. If a person drinks enough to dilute the level of sodium in their blood, this causes the cells of the brain to swell. In the process, water intoxication can mimic psychosis or bipolar disorder.

The Irish doctors Melissa Gill and MacDara McCauley recounted one such perplexing case in a medical journal, sharing the cautionary tale that they deemed "catastrophic" in hopes of saving future doctors from making the same mistake. Against his will, an alcoholic, bipolar forty-three-year-old man was brought to their hospital after "engaging in uncharacteristic behaviors such as blowing smoke in his son's face and kicking the family pet." He told them he was paranoid that people were talking about him, and was lacking in energy and having difficulty concentrating. In the doctors' report of the case, they note that he "displayed poor insight" and "appeared perplexed."

Perplexed themselves, Gill and McCauley prescribed the antipsychotic medication risperidone—in addition to the nortriptyline and zopiclone that the man was already taking for bipolar disorder. But they went on to note that his condition "deteriorated significantly." He refused to bathe and "exposed his genitals on a number of occasions." The medical team moved him into isolation and renovated his medication regimen. Yet his decline continued.

Finally, a nurse noted that the man had been drinking quite a lot of water—an observation that would have been easy to dismiss had the case not been so difficult to solve. So the medical team checked his sodium levels. Predictably, they were low—he had mild hyponatremia.

He continued to get worse, exposing himself to staff and other patients, and urinating publicly on the ward, saying that God had instructed him to do so. And then he had a tonic-clonic seizure—the most common symptom of hyponatremia that brings people to hospitals.

An emergency CT scan showed that the man's brain was swollen. And in short order his sodium level fell to a critical low. The lower limit of normal is usually drawn at 130, and the man's was 108.

Rushing the man to the intensive care unit, Gill and McCauley set about the critical work of regulating the man's water intake. In this delicate state, they had to take extreme caution not to let his sodium level increase too quickly. Rapid changes can be fatal, destroying cells in the brain stem.

As his sodium level gradually returned to normal, he regained coherence. There was no more masturbating. He was able to go home, where he remained free of seizures.

The case is an extreme example of what can go wrong when our sodium and water levels are off. No one knows exactly what causes psychogenic polydipsia, but one hypothesis is that psychological stress and acute psychosis can reset the body's "osmostat"—a conceptual thermostat for the sodium in our blood. Others believe that thirst is related to dopamine levels, which are altered by many medications. Antipsychotic drugs (like olanzapine) can, meanwhile, dispose a person to seizures.

Of course, few cases are this extreme. Susan Yeargin and other experts are clear that water alone is sufficient to hydrate people in most circumstances, as long as they eat food at least occasionally (and as long as that food contains some salt). Our kidneys will keep the body's sodium concentration high enough by releasing concentrated (dark) or diluted (clear) urine.

Others are more upset about people blindly endorsing water alone as a means to hydration. The physician Eduardo Dolhun is informed by experience with disaster zones and life-threatening dehydration. He has come to believe that subtle cases of water overdosing are all around us—especially when people exercise for prolonged periods or stay out on hot days, sweating out water but drinking enormous bottles of water.

"If people would stop drinking so much water, everything would be balanced," he insists. "Do you see the Maasai running with water bottles?"

Dolhun is a family physician in San Francisco, where he helps direct Stanford University's Ethnicity and Medicine course, which educates medical students on the role of race and culture in caring for patients. He describes himself as a humanitarian and a disaster relief expert. When he talks about hydration, his "life mission," his voice quakes: "Eight glasses of water a day was pulled from someone's butt."

That's tame compared to what you hear from Dolhun if you say the words "Smartwater" or "Gatorade."

So I need a "sports drink"?

In medical school at the Mayo Clinic in 1993, Eduardo Dolhun went on a volunteer trip to Guatemala. There he found himself in the middle of an epidemic that had ascended from Colombia, an explosive bacterial disease that kills one hundred thousand people every year: cholera. "It's the fastest known way to dehydrate a human body," Dolhun said.

Cholera is the mechanism on which all dehydration standards are based. The disease is straightforward: Cholera kills by dehydrating. It does not damage the colon structurally, but simply turns its fluid channels permanently to the "on" position until the body is empty.

If a person can manage to stay hydrated, though, the infection will pass within a few days. This has proven one of the most difficult challenges in history, testing the limits of our bodily knowledge throughout centuries of gruesome epidemics. Today it is only because of the research that went into understanding cholera that *electrolyte* is a buzzword in SoulCycle studios. Cholera research is the reason that Gatorade has some scientific basis for putting sugar and salt in water. So even if only to understand what is best to drink during hot yoga in Wicker Park or CrossFit in Oakland, it makes sense to start by understanding cholera.

The disease divides the world along a clear line. On one side are the places that have clean water, where cholera affects no one and seems archaic, where its mention is usually a kitschy reference to the 1971 computer game *The Oregon Trail*. On the other side are places where clean water is not reliable, and where cholera collectively affects millions every year. There a person can become infected and grow so quickly dehydrated that within hours their eyes will have sunk into their head and their skin wrinkled, developing what patriarchal medical texts call "washerwoman's fingers." In a few hours to a few days without treatment, half of these people will die. One refugee camp in Rwanda during the genocide of 1994 saw the mortality rate from cholera diarrhea reach 48 percent, with more than twelve thousand Rwandan Hutus dying of dehydration. Almost all of them could have been saved with proper oral hydration.

In Dolhun's experience as a first responder in Haiti, the Philippines, and Nepal, dehydrated patients who drink pure water will only hasten their deaths. He sees Gatorade as no better. "Doctors giving Gatorade to anyone with [clinical] dehydration should be malpractice," he implored. "Okay? Malpractice."

Susan Yeargin advises athletes who are disposed to heatstroke in the South Carolina summer that too much Gatorade can be just as dangerous as too much water. The sodium in the mix is too low, so "those people would still be at risk for hyponatremia if they had been chugging Gatorade."

Cholera has killed tens of millions of people in the last two hundred years, since advances in transportation and global trade allowed the ancient disease of the Ganges delta to escape the Indian subcontinent and spread around the world. No one knew the cause until a cholera pandemic hit London in the early 1850s and a doctor named Jack Snow began plotting where his patients were dying—and where they got their water. Overlaying the maps, Snow deduced that it was a waterborne infectious agent. Germ theory was still gaining steam, so his assertion that a microbe could be to blame for an epidemic was met with stern resistance from germ-theory denialists.

Snow was proved correct three decades later, though, when microbiologist Robert Koch—fresh off of his discoveries of the bacteria that cause anthrax and tuberculosis, and on the verge of discovering those behind gonorrhea and syphilis—identified the bacterium *Vibrio cholera*. Our acidic stomachs protect us by killing most disease-causing bacteria, as does the mucous lining in our bowels. But *V. cholera* evolved to have an enormous tail-like flagellum that sweeps back and forth, propelling the organism onto the cells that line our bowels, enterocytes, releasing a toxin that causes them to purge water and sodium at a rate of up to two liters per hour. Patients were made to lie supine on cots with holes cut in them and buckets underneath. The treatment was to hydrate these people and wait. If the treatment was done improperly, they died.

The discovery in the 1920s that saline fluid could be delivered through a needle directly into a person's veins—intravenous (IV) hydration—was a boon for cholera treatment. Saline tastes like blood by design. Less salty than ocean water, but not like something you'd want to drink, saline is meant to match our blood's concentration of sodium. As Dolhun put it, "IV fluid was essentially a salt delivery mechanism."

But it is not with haste that one should endeavor to inject anything into human veins. The process requires calculation, sterile production facilities and syringes, and trained administration. This was, in many places ravaged by cholera, not a practical solution. As recently as 1982, cholera was the main cause of infectious diarrhea that led to the deaths of around five million children under the age of five every year.

But ten years later, that number was down to three million. By July 2012, the World Health Organization estimated that cholera killed only 120,000 people. So what caused the enormous reduction in just three decades?

The "magic bullet," in the words of Columbia University's Joshua Ruxin, was something called *oral rehydration therapy*. Which is to say, drinking. But obviously not water or Gatorade. So—what?

Initial attempts to treat people with cholera by giving them pure water only led them to die more quickly because it precipitated lethally low concentrations of sodium in the blood (as happened with the bipolar Irish man). From this, we learned that hydration is all about concentrations of sugar and electrolytes. Keeping them all in balance depends on a force of physics, diffusion—the principle that if you mix a concentrated solution and water, the result will be a diluted solution. So, for example, pour a shot of coffee into a glass of wine, and the result is coffee wine. (Without diffusion, it would just be a clump of coffee adrift in a glass of wine. Difficult to imagine.)

"The big mistake in oral hydration is everyone tries to put more electrolytes in," William Greenough, a professor of medicine at Johns Hopkins, told me. If you pour a concentrated elec-

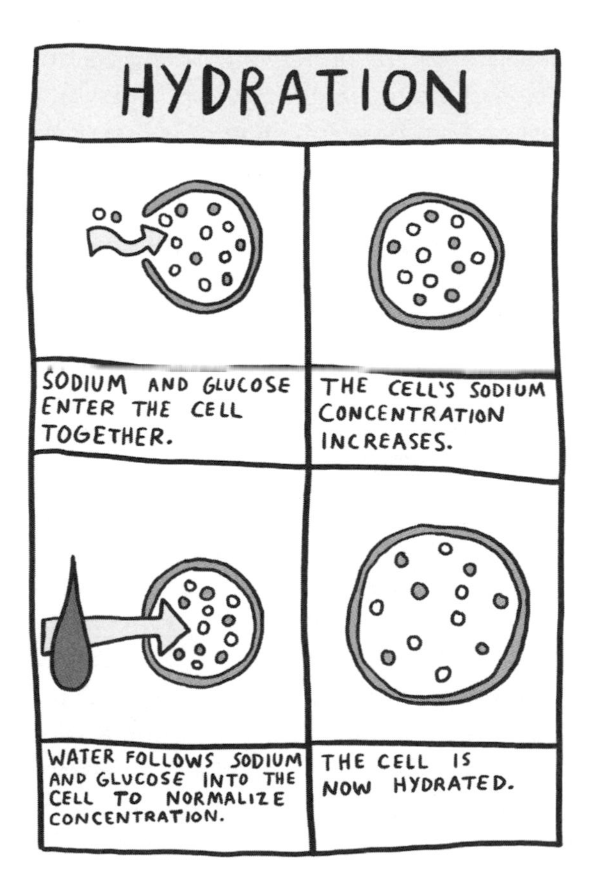

trolyte solution into your bowels, diffusion will draw water out of the body and into the bowels until the concentration is lower. The critical thing to remember, Greenough emphasized, is that this diffusion happens because "the gut is a very leaky membrane." Pouring a shot of coffee into your body is just like pouring one into a glass of wine. Components will equilibrate, moving in both directions across the bowel wall.

This is how "sports drinks" can dehydrate us. In normal conditions, inside the cells of our bodies there is salt-sugar water at low concentrations. When we drink something with higher concentrations, diffusion pulls water out of those cells and *into* the bowels to equalize the concentrations. So even though you are drinking fluid, it can be effectively pulling water out of you.

The key to hydration came fifty years ago with a seemingly simple discovery. Using the intestine of a literal guinea pig, in 1958, Canadian sugar physiologists discovered that glucose could not cross the membrane of the bowel by itself. It had to travel as a pair, and its partner was sodium. This inspired another sugar physiologist, the American Robert Crane, to define the mechanism: He discovered a series of tiny gateways in the cells of the bowel that transfer sodium and glucose, simultaneously, in and out of the body's cells. Crane's discovery of the sodium-glucose transport pump, a tiny gateway into the cells of our guts, revolutionized hydration. A 1978 editorial in *The Lancet* would call this "potentially the most important medical advance this century."

Of course, the textbook histories of many medical discoveries are predicated on some benevolent intellectual deigning to share his (usually *his*) revelation with the suffering people of the less developed countries. These men are painted as heroes. This savior narrative tends to falsely presume that much of the world is in need of saving by white men. When people are genuinely in need, the savior narrative usually overlooks or ignores the cause of that suffering. Namely, in this case, why were so many people dying of dehydration?

Why do so many people die of dehydration?

At the signing of the Treaty of Paris at the close of the Spanish-American War in 1898, the United States took possession of the Philippines from Spain for a price of $20 million. The Philippine people were not consulted; earlier that year, they had declared independence.

Just months before the treaty, Philippine revolutionaries had fought alongside American forces to expel Spain from the islands—apparently believing that Americans were their liberators. The Philippine resistance had led to the freeing of Manila in August. So they were surprised when American forces rolled in to expel them. On the evening of February 4, 1899, two American sentries in Manila fired on a Philippine man, and tensions erupted into the Battle of Manila.

The following day, attempts at diplomacy were rejected by the American military governor, General Elwell Otis, who replied that the battle must go on "to the grim end." And so over the course of four years, Americans waged war on the people of the Philippines. They drove civilians into concentrated pockets where living conditions were deplorable, and predictably, cholera broke out, killing approximately two hundred thousand people—many more than were killed in battle. Historian David Silbey describes the scene as apocalyptic, the country "collapsing into barbarity and chaos," culminating as the cholera epidemic was accompanied by an infestation of literal locusts.

After decades of neglect of its colony, the United States eventually agreed to the formation of the Philippine Commonwealth, which was to transition to a fully independent state. But World War II derailed that plan. Instead of independence, within hours of Japan's bombing of Pearl Harbor the Philippines came under attack from the island empire as well. Japan swiftly dispensed with American forces and occupied the country.

This brutal period included the Bataan Death March, among

seventy-two war crimes committed by the Japanese, approximately a million murdered Philippine people, and further deterioration of the country. When the war ended with the Treaty of Manila, the nation finally achieved independence, bereft of infrastructure. So in 1961 a new, especially dangerous strain of cholera broke out.

And that is why valiant Americans needed to come to the rescue with their oral rehydration therapy.

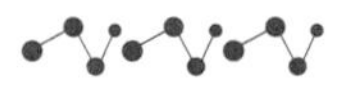

The U.S. Navy transferred Dr. Robert Phillips from Taipei to Manila, where he set up a cholera clinic at San Lazaro Hospital. For years in Taiwan and elsewhere, Phillips had been forced to treat people without IV hydration, and he had been experimenting with oral hydration therapy. He knew that keeping a person's sodium from dropping too low was critical. The problem was that even when he put sodium into water, people didn't seem to absorb it. They still died. It was in Manila that he implemented Crane's discovery and added glucose to his electrolyte solutions. That's when hydration changed forever, and "the most important medical advance this century" earned its title.

In August 1962, Phillips's team treated three patients with the oral electrolyte solution of glucose and sodium, and all three recovered. This appeared to be the first proof of the effectiveness of sodium-glucose absorption in humans. The team set up a clinical trial to test the new solution on more people. The attempt was hasty, but without IV saline, the alternative was death.

The only way to really treat cholera would be if oral hydration alone could be given in the field. And by the thousands—in people's homes, by friends and family and anyone else who was not too ill to mix glucose, salt, and water.

Physicians David Nalin and Richard Cash undertook the first recorded treatment of cholera patients using a glucose-sodium solution alone, given exclusively and entirely by mouth, during April 1968. In August, their results were published in *The Lancet*.

But as a cure for an epidemic, the treatment was still hypothetical until in 1971 Pakistan's army attacked East Pakistan, chasing nine million people over the border into India. This sparked another cholera outbreak in refugee camps along the border. There a young William Greenough, then a Harvard resident physician, and a handful of other doctors in Calcutta attempted to stem the spread of the disease. The team trekked with thirty liters of IV saline to a camp along the Shitalakshya River, near the Lakshmi-Narayanganj jute mill. In the small compound, they found five thousand people. They ran out of IV fluid in about thirty minutes.

"The only recourse then was to separate the sick from the well," Greenough remembers, an especially difficult task among families during a time of war. The mortality rate among the refugees was around 40 percent.

The desperate physicians, though, caught word of oral rehydration therapy. Greenough said at the time that, given the amount of fluid a person loses in cholera, oral rehydration "was quite an outlandish idea." But, confronted with little choice, Greenough and colleague Norbert Hirschhorn attempted to implement and tweak Phillips's solution. They employed it only in dire cases, on people who arrived in shock, and only as an addition to what little IV fluids were available. Hirschhorn was careful to be sure that the oral solution was not more concentrated or less concentrated than what is inside human cells. That winter, the team showed that in every case, the sodium-glucose solution caused people to absorb and retain the fluid. So they used it more and more. Within one year, the mortality rate of cholera in the Dhaka hospital was under 1 percent.

In the refugee camps, Greenough's team cut the mortality rate from around 40 percent down to 3 percent with no IVs at all. It worked even in severe cases, when people were so dehydrated that they had gone into shock.

That caught the attention of the World Health Organization and UNICEF, which launched major global initiatives to supply oral rehydration solution around the world. "Getting from basic science to practical use took place with breathtaking speed," Greenough recalls. It went from field use in 1964 to global programs in less than a decade. "Once people discovered that they could drink this solution and not die—even though it doesn't taste that great—it was taken up quite rapidly."

By the 1980s, largely due to this simple solution, the mortality rate of children under four had fallen from around eight million to two million per year. It's now under one million.

And yet in much of the wealthy world, there was silence. There still is. Not only are most people unfamiliar with oral rehydration solution, but doctors almost always prescribe IV hydration instead. Greenough, Dolhun, Ruxin, and others are blunt as to why.

"Quite simple," Greenough puts it. "If you come to the emergency room with dehydration, and I sit you down and give you a drink, I can't bill you for it. For IVs, the people who make the fluid make a lot of money, the people who make the needles make a lot of money, the people who make the tubes make a lot of money, the hospitals make money, and the physician makes money. Everything is against it."

Ruxin surmises that the discovery of oral hydration illuminated how "prejudices of the medical establishment, and its reverence for advanced technology, can postpone lifesaving discoveries." Or as Hirschhorn said of oral rehydration therapy on the BBC, "Its simplicity was its own enemy."

"It's not interesting on TV, or in the media, to see someone taking a drink," said Greenough. "Go to the ER, and if your blood pressure is low, people rush around, you get an IV, a CT scan while you're in shock—now, this is exciting stuff. But there's no need for it."

As an attending physician at Johns Hopkins Bayview Medical

Center, where he still practices today, Greenough estimates that around 15 percent of people admitted to the hospital could have stayed out with proper oral hydration.

Unless you're pushing your body to its physical limits and going a long time without food, say in an Ironman Triathlon, drinking a sodium-glucose solution instead of water is unlikely to mitigate the extra calories and arguably harmful addition to daily sodium totals. Except in cases of severe dehydration, water should work. Drink water and eat food that contains salt and carbohydrates, and you have your own oral rehydration solution factory inside you.

The real importance of this science is that dehydration from diarrhea is still the number two cause of preventable death around the world for children under five. A person with severe cholera cannot survive by drinking water, sterile or otherwise. In wealthy countries, the solution is to keep those people hydrated through needles in their veins, out of tradition rather than logic. Where IV fluids—and people who know how to administer them—are in short supply, packets of sodium, potassium, and glucose have proven to be what Doctors Without Borders has called "the most important medical advance since penicillin."

Oral rehydration solution is in need of champions, but it's also important as the standard by which to understand what we put into our bodies. Gatorade is culturally normal; oral rehydration solution like the one that most people know by name—Pedialyte—is not. Throughout my own trials of oral rehydration solutions—I tried out every one that is commercially available over the hottest summer months in New York—not one person that I told about it said anything like, "Oh, that sounds reasonable."

It may be because, when mixed in the right quantities, sugar-salt-water solution is classified as a "medical food" by the FDA. When you buy Pedialyte in a pharmacy—for around $6 per liter—it comes sealed in plastic, like cough syrup. A couple aisles over, the nonmedical beverages are the twist of a cap away. If any food is considered a medicinal product—which Pedialyte is—is it odd that all the rest of our food isn't?

When I was having dinner with a psychiatrist friend, she laughed as I told her about my experiments with oral rehydration solutions, saying that I was always "sort of Asperger-y." (*Because why else would someone question a paradigm?*, I shouted, standing and upending the table.)* More important was her admission that she, a medical doctor, had never heard of oral rehydration solution. Handing out packets of sugar and salt to be mixed in water does not require the intensive training which doctors pride ourselves on completing. It is, rather, at odds with the process.

"It's looked at as a poor-country remedy," said Greenough. "This is a beautiful mechanism that the body has developed over millennia. . . . But at some point, when we get tired of spending $3 trillion on our health, we will use it."

What about Smartwater?

Eduardo Dolhun sounds like his tongue is being pierced as the word *Smartwater* leaves it. In terms of bottled water sales in convenience stores, Smartwater leads all competitors, with $350 million in 2015, despite being among the most expensive bottled waters on the market. Why?

"Someone came up with this unbelievably great name, Smartwater," Dolhun said.

Someone also put images of Jennifer Aniston drinking it all around us, often nude. The ads claim that the water is "electrolyte enhanced," despite the fact that it contains no sodium.

"Electrolyte enhanced?" Dolhun shouts. "What does that mean? It means bullshit."

Smartwater, then, is water with trace amounts of calcium

* I didn't do that.

chloride, magnesium chloride, and potassium bicarbonate. The small print notes that these are added "for taste." The sodium content is zero. A similar product in the Whole Foods 365 line, called Electrolyte Water, is marketed with the claim that "Proper hydration is crucial to overall wellness!" Which is true, but the implication that the electrolytes in these products constitute proper hydration is misleading. As Dolhun put it, "The city water in Philadelphia has more electrolytes."

Indeed, much tap water contains roughly twice as much of those elements as does Smartwater. When there are too many electrolytes in tap water, it develops some opacity, and people complain. But Coca-Cola (or Glacéau, its privately owned subsidiary and manufacturer of Smartwater) can sell it as "electrolyte enhanced" because it is not technically false.

"Would you undergo chemotherapy with a 'chemotherapy-enhanced' IV?" Dolhun posits. He begins an internal dialogue, presumably between a doctor and a patient:

"What does that mean?"

"Oh, I don't know, we're just going to give you some [highly toxic cancer-killing drug] cisplatin."

"But at what level?"

"Just trust us, we'll just give you something."

When you have watched people die of dehydration, maybe his scenario seems less absurd by comparison. For most of us, it's just a waste of money and, as with any bottled water, a punch in the gut of the environment. Ads boast that Smartwater is "vapor distilled," a process that is "inspired by the way nature purifies water."

So it's boiled?

Coca-Cola clarifies that vapor-distilled means "heat is used to vaporize water"—okay, yes, boiled—before "the vapor is then cooled and the water condenses back to a purified state." This is the water cycle—like diffusion, another middle school science concept that turns out to be worth knowing, lest it be used to take your money. The explanation omits the fact that, as opposed to simply filtering water, boiling it is an energy-intensive process. So nature may have *inspired* it, but it is no more friendly to nature than the plastic bottles in which this boiled water lives.

Juice is healthy?

Juicing fruits and vegetables is as logical as juicing any worldly possession. Imagine how many more shirts you could fit into your closet if you juice them. Juice your car and suddenly you're the guy with forty cars in a two-car garage.

Juice lives in a privileged place where it can find favor even with people who hate "processed" foods. Dariush Mozaffarian, the dean of the Tufts University Friedman School of Nutrition Science and Policy, warns against blanketly condemning all such foods, even as his own work has shown that 58 percent of calories eaten in the United States come from products that qualify as "ultra-processed"—and that this is definitively contributing to metabolic disease. "Processed" is too broad a term, though, and not practical for the world where almost all food is processed. Though there clearly is something to the concept.

Juicing stands apart as a singular form of processing: the stripping of fiber from fruits and vegetables. If I had only one number to look at on a food label, and from it guess at the health-promoting qualities of that food, it would be fiber. Purposely removing fiber from a food is curious, and is one of the few acts in the realm of nutrition that might be definitively said to be bad.

Juicing makes sense from a twentieth-century reductionist approach to nutrition. In the first half of the 1900s, we had just discovered vitamins and come out of an economic depression where the idea of having *too much* was all but unheard of. If an apple a day is good, the thinking went, a hundred apples a day must be better.* If you make juice, you get more vitamins into less space. You get the vitamin equivalent of twenty oranges in the space of one orange. Let's give it to the astronauts.

It's easy to criticize post–Great Depression approaches to health. I haven't lived through times when the idea of "too many" nutrients was a fantasy, so I shouldn't be too critical. But here I go:

Pouring liquid sugar (juice) into your stomach sends a high dose into your liver and blood. The pancreas races to secrete insulin to lower your blood sugar. Then the sugar load is over, and the insulin is still there, so your blood sugar gets low. You feel bad and crave something sweet. We can and do easily become obese and diabetic from drinking too much juice. Barry Popkin, a distinguished professor of nutrition at the University of North Carolina, now devotes himself largely to juice awareness. He is among the many who have told me they see juice to be just as bad as soda. Many of both types of drinks are produced by the same companies. Coca-Cola acquired Odwalla in 2001, and PepsiCo acquired Naked Juice in 2007.

Because the vilification of carbohydrates is not going anywhere quickly, Dariush Mozaffarian and his colleagues at Tufts devised a rule of thumb for differentiating "good carbs" from "bad carbs." His proposal is a step forward to promote at least some distinction within the category now largely referred to reductively as "carbs."

Look at the number in the "carbohydrate" line and compare it to the number in the "fiber" line. In a "high-quality carbohy-

---

* Researchers at the University of Michigan have investigated whether people who eat apples every day are less likely to see doctors, and they found no correlation. So, they found, an apple a day does not keep the doctor away. The unsettling study was part of the "April Fools Day" issue of the medical journal *JAMA Internal Medicine*. But the study was real. This is why we need studies of humor. And the whole thing gets at the question: Is it *good* to keep the doctor away? Sometimes the sickest people see the doctor the least, partly because they don't have good access to health care, partly because of cultural aversions rooted in oft-brutal historic precedent, and partly because it can be easiest to live in denial of the fact that we are sick and need help.

drate product," fiber will be at least 20 percent of the carbohydrate content.

The method will not reach many people, because it involves picking up food packages and flipping them to look at their nutrition labels. Worse, there is math.

Ideally the fiber is innate to the food, as opposed to a supplement in a mix (like those gummy bears that have fiber powder in them, which would, unfortunately, qualify them by these numerical terms as "high quality"). After playing around with this rule, it was apparent that it basically rules out most things that come in packages.

Which leaves us back at recommending people eat whole-plant foods, again.

The long-troubling program Weight Watchers recently made a sophisticated move, lowering the point value of fruits to zero. Under their plan, points are bad. So people can eat as much fruit as they like. Yale's David Katz has several times challenged me to show him a person who is obese from eating too many whole fruits and vegetables. Eating twenty apples is tough because the fiber fills your stomach. Whole fruits and vegetables will take you longer to eat, giving your body time to tell your brain that you're full. Even if you can do it, combine the time it would take to eat all twenty of those apples with the effect of the fiber slowing the absorption of sugars in your bowels, and the effect on your blood sugar would be totally different from drinking a cup of apple juice.

As juice people have begun to appreciate that "added sugars" are to be avoided, Welch's grape juice has, like Odwalla and Naked Juice, focused its recent marketing on the fact that their juice has "no added sugars." It's kind of like labeling peanut butter with "no added peanuts." Welch's grape juice has just as much sugar per ounce as Coca-Cola. It would be hard to add more without creating a thick, syrupy product (like Welch's jelly).

To expand on Mozaffarian's rule, mine is that if you're drinking something and you have to keep swatting hummingbirds from drinking out of your cup, it doesn't matter how much of the sugar was "added" and how much was extracted from grapes.

Also, just let the hummingbirds drink. They can't store energy like people can, so they *need* to constantly be consuming sugar. You try flapping your arms a thousand times per minute.

Why is there Vitaminwater?

The entrepreneur J. Darius Bikoff was nearing bankruptcy and feeling "run down" one day in 1996 when he did the American thing and reached for a vitamin C supplement. As he recounts of his breakthrough moment, he was also drinking mineral water at the time. And so, like the inventor of the Pizza Hut Taco Bell before him, Bikoff thought, "What if I combine these things?"

Bikoff started a company called Glacéau (also known as Energy Brands). The name conjures glaciers, but the source was Connecticut groundwater. His product, Vitaminwater, was the result of sugar, coloring agents, and various isolated vitamin compounds. It hit the market in 2000. Its various formulas were called Revive, Power-C, Energy, Focus, and Essential. The motto, "vitamins + water = all you need." (It made no mention of the sugar or coloring.)

As people begin to accept that water is more prudent than soda or juice, the fastest-growing part of the beverage industry has become a fascinating line of products known as "enhanced waters." The International Bottled Water Association defined these as waters "with added fluoride, essences, or supplements." That has come to include sugar. Most tap water has been purified to remove minerals and microbes, and much of it has been fortified with fluoride. But this is not *enhanced* water, so it is a question for philosophers: At what point does water become enhanced? And more importantly, at what point does that enhanced water become not water, but soda or juice?

Vitaminwater, now made by Coca-Cola, contains 33 grams of sugar per bottle. This is just less than a can of Coke, which has 39 grams. Which could be confusing, because the word "water" is right there in the name.

Since bottled water was introduced in 1977, the bottled water industry has grown to $11.8 billion annually in the United States alone. For the first three decades, this product was clearly distinct from bottled soda, bottled juice, bottled milk, and so on.

In 2003, Glacéau employed two hundred "hydrologists" to canvass the country and educate people on the health benefits of Vitaminwater. But their physical reach was necessarily limited. It was difficult to break out of the usual expensive-water-drinking customer base. What they needed was someone who could lead people blindly, in large groups. It so happened that the same year Vitaminwater hit shelves, a Junior Olympic boxer turned crack dealer turned amateur rapper named Curtis Jackson was shot multiple times while standing in his grandmother's front yard in Queens. He survived and, under the name 50 Cent, devoted his life to what would become the title of his debut strudio album three years later, *Get Rich or Die Tryin'*. Fortunately for Glacéau, that mantra would include an openness to partnerships with purveyors of enhanced waters.

When the single "In Da Club" topped the *Billboard* charts in 2003, the world's overnight allegiance to the fitness-conscious rapper caught the eye of Vitaminwater marketing executive Rohan Oza. The two met and agreed to work on a product that would appeal to 50 Cent's fan base. The rapper had grown up drinking 25-cent "quarter waters" from corner bodegas in Queens. A sort of generic Kool-Aid, usually dispensed in plastic bottles shaped like barrels, quarter waters have a sugar content similar to Vitaminwater and juice. The transition from quarter waters to Vitaminwater was an easy one for Jackson, who would later say he was drawn to the company because "they do such a good job making water taste good." With quarter water, his flavor of choice had always been grape. So the team settled on grape as the flavor of the new concoction: Vitaminwater Formula 50.

In exchange for ongoing endorsement of Vitaminwater, 50 Cent received a significant stake in Glacéau. Soon he was pounding Vitaminwater on billboards and bus-stop posters in neighborhoods previously untapped by luxury waters. An embodiment of brand crossover, in one commercial Jackson conducts a full symphony orchestra from Beethoven's Ninth into an arrangement of "In Da Club." A purple bottle of Formula 50 rests atop his podium. In a subsequent performance at the BET Awards, Jackson concluded a song by raising a fist in the air and delivering the parting words "Vitaminwater, ladies and gentlemen, Vitaminwater."

In 2006, Glacéau founder Bikoff purchased a four-thousand-square-foot penthouse in Manhattan for $5.6 million. In 2007, he sold Glacéau to Coca-Cola for $4.1 billion. That deal set a record in hip-hop history as well, with the *Washington Post* and *Forbes* reporting that 50 Cent cleared $100 million. That summer, he released a song called "I Get Money," in which he rapped, "I took quarter water, sold it in bottles for two bucks; Coca-Cola came and bought it for billions, what the fuck?"

This was very much the point of the Center for Science in the Public Interest when it brought a class-action lawsuit against Coca-Cola over deceptive marketing practices. Over the course of seven years, the consumer advocacy group sued Glacéau over its Vitaminwater, arguing that the name and marketing of this particular sugar-delivery product was misleading.

Mike Jacobson, director of the center, sounded exhausted when he recounted the lawsuit to me. "They were making claims like 'rescue,' 'energy,' 'focus,' 'revive.' And it's really just sugar water. With ordinary vitamins. They don't revitalize you. They don't revive you or help you focus. It's all phony." A highlight for him was several years ago when Coca-Cola's attorneys argued that "no consumer could reasonably be misled into thinking Vitaminwater was a healthy beverage."

The judge did find, in 2010, that it was a violation of FDA rules for Vitaminwater to specifically use the word "healthy." Ads calling Vitaminwater "nutritious" were found to be "misleading" and banned in the UK. Glacéau has been forced to step back from certain other words where legally mandated, and has been required to include on Vitaminwater labels the words "with sweeteners." The company continues to use labeling that carefully implies a functional effect without overtly stating it. The latest iterations are called Tranquilo, Connect, Spark, and Stur-D. The producers have also gone on the offensive, suing PepsiCo for infringement for the packaging of its sugar drink SoBe as "Life Water."

Coca-Cola also sells a product called Fruitwater that, they admit on their site, "by design . . . contains no fruit or fruit juice." That's a little more defensible than "Oops, no fruit or fruit juice," but it still makes you wonder, why call it Fruitwater? I asked Jacobson if he had considered suing over the use of the term "water." He said that's tougher to define. Technically water is the primary ingredient in these juices and sodas—as it is in beer and coffee.

Or, you know, Barley Water and Bean Water. Don't drink the Oil Water. It's for cars.

Is drinking seltzer the same as drinking regular water?

The ascent of seltzer water in the last few years has been unprecedented, even in the long history of seltzer. Detractors abound, whispering that seltzer is bad for our teeth or our bones because it is acidic. In the interest of public health, we might be better off if these people had no teeth or bones themselves, and so could not spit nonsense.

(That's hypothetical. Don't remove anyone's teeth or bones.)

The colorful cans of LaCroix are one fad that has done good for health. The product is not new—I remember my aunt drinking it in northern Wisconsin when I was a child—but LaCroix is now fetishized in Los Angeles and New York, and not with the winking inauthenticity of hipsterism, but the genuine passion rarely roused among cynical young adults. Despite almost no marketing or advertising, sales of seltzer water have doubled in the last five years, with LaCroix leading the way.

(A quick semantic clarification: LaCroix attempts to define itself as a "sparkling water" as opposed to a "seltzer," because it uses only "natural flavors" and has no sodium. In fact, most "seltzer" waters have no sodium, and the idea that suspending any flavoring agent in carbonated water can be "natural" is a misunderstanding of nature. Health equity hinges on acknowledging the labels with which people identify; this right to self-definition does not extend to beverages.)

Even if it's undergirded by capitalism, seltzer is indeed the face of resistance. Its ascent is a paradigm of populism in health. A shift from soda to seltzer stands to improve global health, and not in a trivial way. A can of Coke contains ten teaspoons of table sugar. According to the Harvard School of Public Health, the addition of one daily can of soda or juice to a person's diet will result in around five pounds of weight every year. These are calories that we add to our diets that paradoxically serve to make us hungrier. Eliminating and replacing sugar-heavy drinks with water could leave many people less obese and sick.

And if carbonating that water is an incentive toward that end, it's worthwhile. When you infuse water with carbon dioxide, the result is carbonic acid. It's difficult to deny that this name contains the word "acid." But acid is a word, like "sandwich" or "TV drama," that describes a spectrum. Carbonic is a very weak acid, the *Chicago Hope* of acids. One role of saliva is to help neutralize acids like this and protect the teeth. If you hold sulfuric

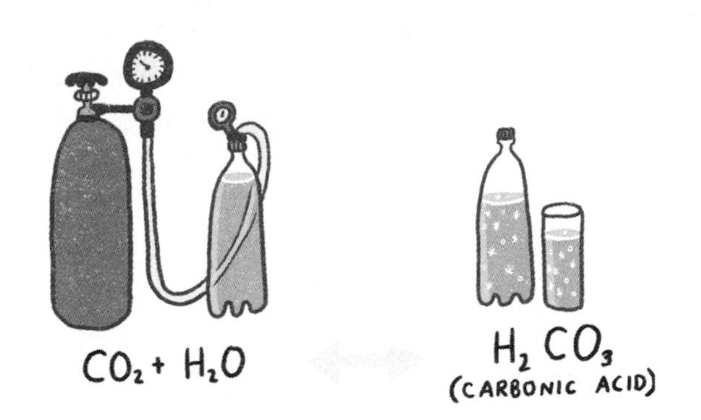

acid in your mouth, your saliva will be rendered useless, as the acid eats through your tongue and palate, causing devastating chemical burns. Its pH is 0.3, on a scale where 0 is the most acidic and 7 is water, which is neutral.

Carbonic acid has a pH of 5.7 in water. A freshly opened can of seltzer will have a slightly lower pH, but still, saliva can handle that.

Chug seltzer and the worst that will happen is the bubbles will get together in your stomach and mutiny, returning to daylight as a burp. But swallowing some air is a normal part of drinking anything. People who burp a lot do so because they swallow more air than do civilized people.

The acids worth focusing on are those not in seltzer, but in soda and juice: phosphoric and citric acids, which have acidic pH values of 1.5 and 2.2. That's closer to sulfuric acid than to carbonic acid. Citric and phosphoric acid give sodas their sharp, tangy flavor. It's what's marketed as "refreshing" when LeBron James takes a gulp of Sprite and then turns to smile at the camera with his regal, milky dentition. But if you carefully review NBA footage, you may notice that James is not seen drinking Sprite during games. It's as though he does not care about refreshment.

The carbonated "sports drink" All Sport enjoyed only a brief existence in the 1990s. Despite heavy marketing of All Sport by PepsiCo—the sort that succeeded in turning soda into a normal thing to drink during *every other daily activity*—human physiology drew the line at carbonated sports drinks. (It now exists in a limited-distribution form that is not carbonated—and does have added B vitamins.)

While the carbonation is disruptive when chugged, it's the phosphoric and citric acids that damage teeth. Left soaking in Coke overnight, a pristine human tooth will be broken down into a gnarled, blackened charcoal-like lump. (I know this because I demonstrated it in my groundbreaking elementary school science fair project.) Since the acid-producing bacterial colonies don't survive long underwater, the experiment isolates the effect of the acidity of the soda, as opposed to the sugar. Seltzer will corrode a tooth less quickly, and without the blackening.

Maybe you're wondering, where did I get the teeth? It's a reasonable question. One for another book.

As concerns about acidic drinks grew, so grew a reaction—a market for a product that is the opposite of acidic: alkaline. Drinking so-called "alkalinized" or "alkaline" water, though, will not change the acidity or alkalinity of your body. In your blood, pH is tightly regulated at around 7.4. Minor fluctuations, on the order of 0.2 in either direction, are seen in serious illness. Below 6.9 or above 7.9 is fatal. So it's actually good that you can't change the acidity of your blood by drinking "alkalinized water."

Our kidneys do most of the work in keeping our pH game tight. In cases when our kidneys can't keep up, like when we become extremely dehydrated (as from diarrhea) or in cases of enormous bicarbonate ingestion, our brain stem will sense that our blood is growing alkaline. It will signal our breathing to slow, causing us to retain carbon dioxide, which will acidify the blood and help restore our normal pH. It's beautiful.

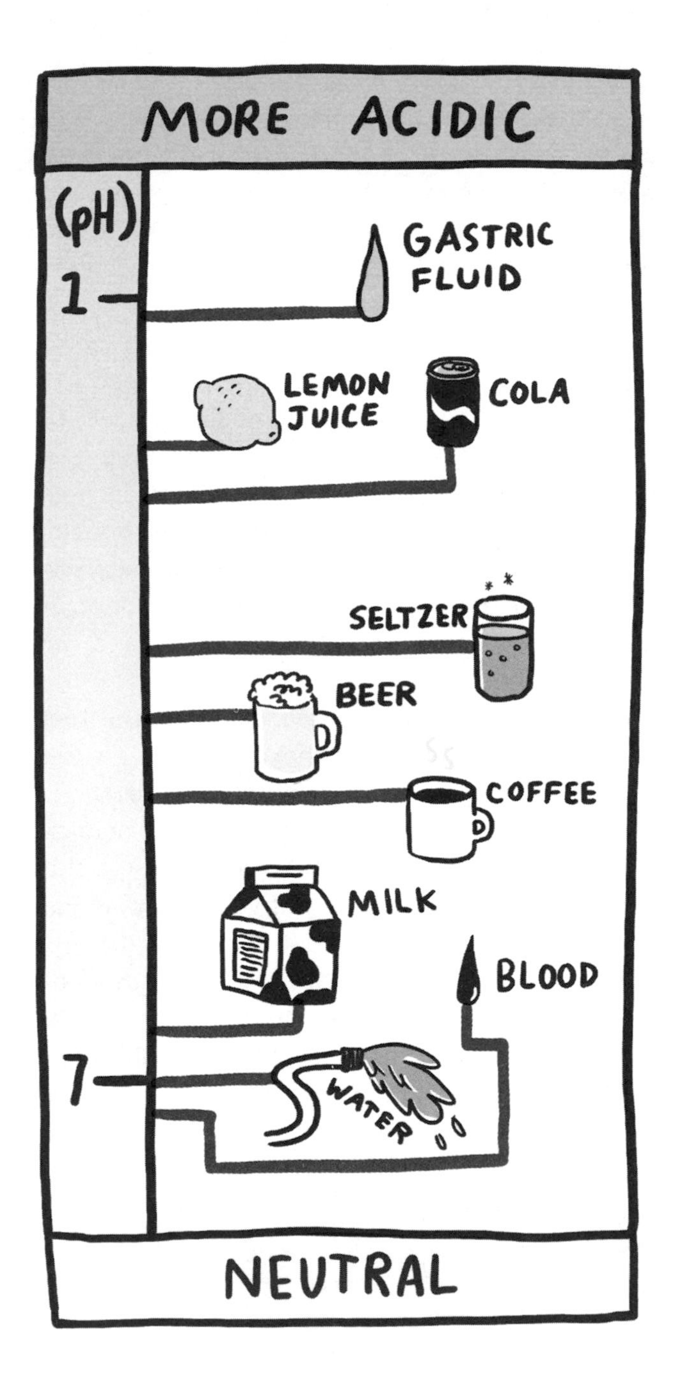

MORE ACIDIC
(pH)
1
GASTRIC FLUID
LEMON JUICE
COLA
SELTZER
BEER
COFFEE
MILK
BLOOD
7
WATER
NEUTRAL

If we were so malleable that some mild alkaline ingestion could change our bodily pH, then the acidity in soda would be enough to kill us. Rather, our ability to keep our pH around 7.4 is, like our sodium concentrations, among the most amazing examples of human resilience.

If I break down and drink a soda: Brush my teeth after, or before?

Brushing right after drinking an acidic drink (juice or soda) is inadvisable, studies have shown, as the enamel is temporarily weakened by the acid. But at the same time, it's not ideal to leave sugar sitting on your teeth, feeding the microbes that turn it into more acid. So I asked oral microbiologist Gary Borisy what one does in this tense situation.

"Yeah," he said, considering the conundrum. "Well. I would get it off. But I don't know. You'd have to do a proper study to know that."

What's driving decay is the acid—primarily lactic acid—produced by strep bacteria in the mouth. Brushing prevents these colonies from building up (and hardening into plaques). But given the apparent imprudence of blindly obliterating ecosystems, a more ideal method is cutting off the resources that feed these colonies. Strep bacteria thrive on oxygen and sugars. Sugars have forever been a part of our diets, but never in the purified forms and quantities of today. Unless you can figure out a way to deprive your mouth of oxygen, reducing the presence of sugar is the way to go. (Which is to say not drinking juice or soda to begin with.)

"And what about the animal kingdom?" Borisy posited, once I got him rolling. "Are the chimps brushing twice a day? To what extent has it become a problem because of our altered diet and the way we live?"

I told him that my dog had terrible breath.

"Well, what are you feeding it?"

Candy, mainly. No, I wasn't. But my dog also pretty much never brushed.

Avoiding all juice and soda is unrealistic for many people living in the real world, but it's also clearly not advisable to leave your mouth colonies bathing in acids and sugars. The worst thing that can happen is to let this decision paralyze you. You just stand in front of your medicine cabinet, reaching out to grab your toothbrush and then pulling back. And then reaching out and then pulling back. And then it is the next day.

While we are blaming modern society for dental ills, it's important to note that none of these beverages compare to the tooth damage that can be done by our own ("natural") stomach acid. Vomiting frequently, as in people with bulimia, or in people who have severe reflux (gastroesophageal reflux disease), imperils the teeth in short order. Even transient exposure to stomach acid causes transparency along the cutting edges of the teeth, with a yellowish hue leading to discoloration as the enamel disappears. This is known as *perimolysis* (*mylos* = molar, *lysis* = breaking down). Because many people with disordered eating are not underweight, tooth damage can be the first thing that a doctor or family and friends notice in a person who may need help.

How does teeth whitening work?

How we choose to drink is the primary factor in the color of our teeth. The enamel of our teeth is not white, but translucent. As we wear it down with acid over the years of drinking soda and juice, it gets easier and easier for pigments to leach into the "white" hydroxyapatite mineral layer. The most common method to bleach these pigments is the same way that stylists bleach hair: Apply hydrogen peroxide, which oxidizes the pigment. It comes in disposable whitening strips and expensive custom-fitted trays.

It also comes in ultra-cheap bottles, which I occasionally swish around in my mouth. I'm not sure if that's good for the microbial ecosystem in there. It's unlikely to be good. The other common option is whitening toothpastes, which use tiny particles to create friction that mechanically scrubs pigments out of the microscopic crevices in the enamel. That can be counterproductive; it wears down your enamel even further, leaving it only more porous and amenable to staining. (So then we buy more whitening toothpaste.)

It would be easier if everyone decided that yellow teeth were beautiful. In parts of sixteenth-century Europe, the thing was teeth blackening. Every trend comes back eventually, I'm told. Maybe easier to get out ahead of that one.

How does fluoride work?

Fluoride is a mineral found in soil and rock. It works its way into the dense hydroxyapatite meshwork that comprises the enamel of our teeth, making them more resistant to being broken down by acid. Though the mineral is found in some groundwater, it is removed during water purification processes, so the United States adds fluoride to drinking water. The system has been in place since it was first tested among schoolchildren in 1948 in Grand Rapids, Michigan. The study was slated to last fifteen years, but the effect was so clearly beneficial that after just eleven years the study had to be stopped because it was no longer ethical to withhold fluoride from anyone. At a time when dying of a dental abscess was still a real possibility, this discovery was profound and quickly acted upon.

Fluoride is a great example of the relativity of our concerns. Now that losing a child to tooth decay is a scarce concern in wealthy countries, certain parents now worry about fluoride. People of privilege and prosperity can spend their time worrying about the most obscure elements of exposure.

Darker conspiracy theories about fluoridating water also

stem from peremptory objections to public programs. I wonder sometimes how different things might be if, in "life, liberty, and the pursuit of happiness," Jefferson had written "health" instead of "life." Or else, "life, but more than just a beating heart." And what that would mean for public health programs of the sort that people tend to object to based on political ideology.

Why are people lactose intolerant?

"Lactose intolerance" betrays a racial and cultural bias, in that about two-thirds of people in the world are "lactose intolerant." Not too long ago, most everyone was. The term is a matter of perspective.

Lactose is the sugar named because it comes from mammalian lactation secretions. If lactose intolerance is a condition—with a name and everything—that implies that the normal state is for us to be able to consume cow's milk and its by-products with impunity. Lactose *tolerance* is the rarer state, which means that drinking cow's milk is a novelty, and certainly not an imperative.

. . . Fine. Why are some people lactose tolerant?

Thank you. For the majority of history, people couldn't drink cow's milk. Trying to do so would result in the symptoms that some today call "lactose intolerance"—bloating, nausea, diarrhea. Worse still was the danger of catching a wild cow and convincing it to let itself be milked.

Like most mammals, humans could digest milk only in the early stages of life when they were breast-feeding. Lactose is a sugar that is unique to milk, and it's broken down in our stomachs by an enzyme called lact*ase*. That turns the lactose into the smaller sugars glucose and galactose that can be absorbed by the cells that line our bowels.

But once we stop breast-feeding, we stop making lactase. Because why would we?

Then, about seventy-five hundred years ago, a new genetic variant known as the LP (lactase persistence) allele appeared in Hungary, where people had begun keeping domesticated cattle. The people with the variant would continue producing lactase into adulthood. Especially in cold climates, where it was difficult to hunt and forage or farm year round, drinking cow's milk allowed people to survive where they might otherwise have died of malnutrition. This allowed people to thrive in otherwise inhospitable tundra. Those people had sex with one another (possibly while drinking cow's milk). The gene spread. People evolved to be able to "tolerate" lactose later and later in life as they produced more and more lactase.

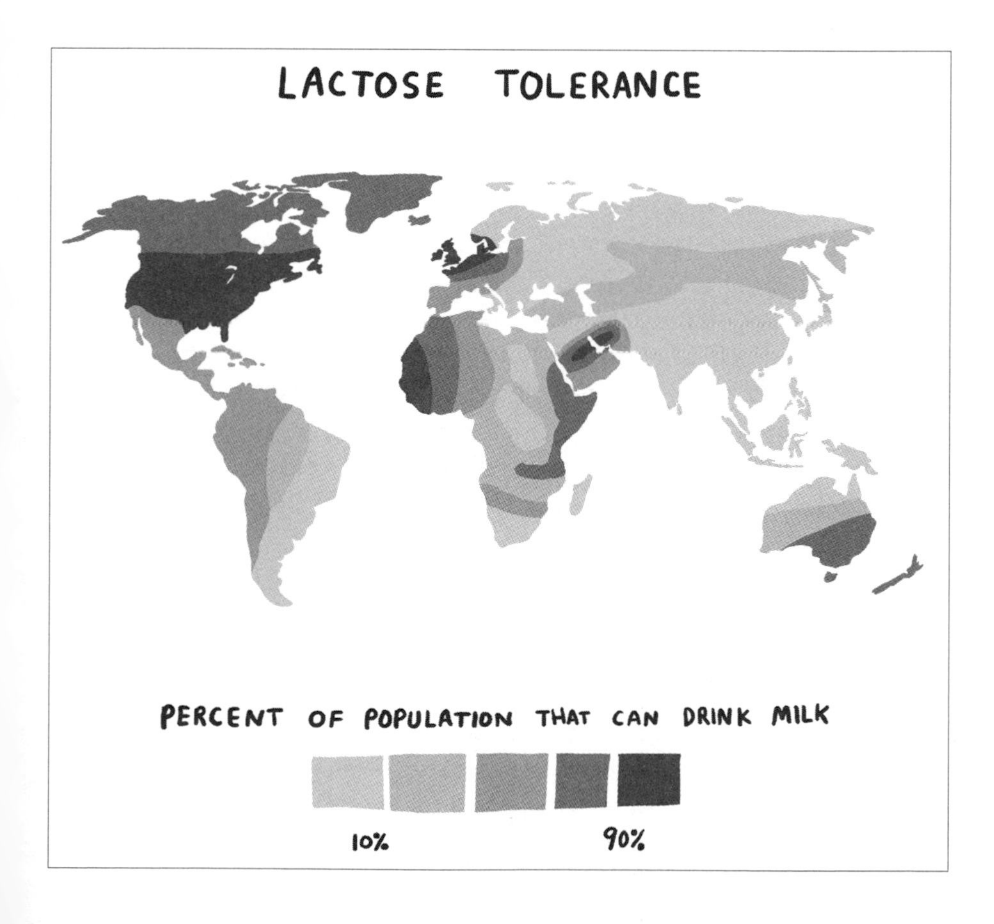

Today, how much lactase we produce—and how late in life we produce it—varies widely between people and between populations. In Sweden, nearly 100 percent of adults can digest lactose. In Botswana, the number is closer to 10 percent.

Does alcohol really kill brain cells?

Alcohol is classified in medical texts as a "neurotoxin." I hate the term "toxin," because everything is a toxin and nothing is a toxin. In extreme cases, water is a neurotoxin. In the trace amounts that it normally occurs in our bodies, formaldehyde is not toxic. As the old adage goes, in alcohol as in all things in life: The dose makes the poison.

In the doses that most people drink alcohol, it doesn't kill brain cells. People who drink heavily will lose some cells. The peripheral nerves are implicated, too, in a condition known as alcoholic neuropathy (which befell the great Hunter S. Thompson).

Especially when it comes to nerves, you don't have to kill a cell to ruin it. The cognitive and emotional deficits that come from alcohol abuse are the result of shrinkage of brain cells and impaired communication between them. That impairment kills brain cells in a real sense. In a bookstore in Cambridge the other day a poster for sale had a picture of Ernest Hemingway and a quote attributed him: "Write drunk, edit sober." Hemingway was profoundly alcohol-dependent and put a shotgun to his head at age sixty-one. Thompson shot himself at sixty-seven. Still, some industrious manufacturer of posters believes the quote should be sellable, not as a cautionary tale, but as an ornamental platitude for the aspiring creative person. One could argue that alcohol effectively killed all of these men's brain cells.

Unlike with a progressive brain disease like Alzheimer's or Parkinson's, the deterioration of an alcoholic person's brain will stop progressing when the person stops drinking. Some recovering alcoholics have been shown to regain brain volume. But that

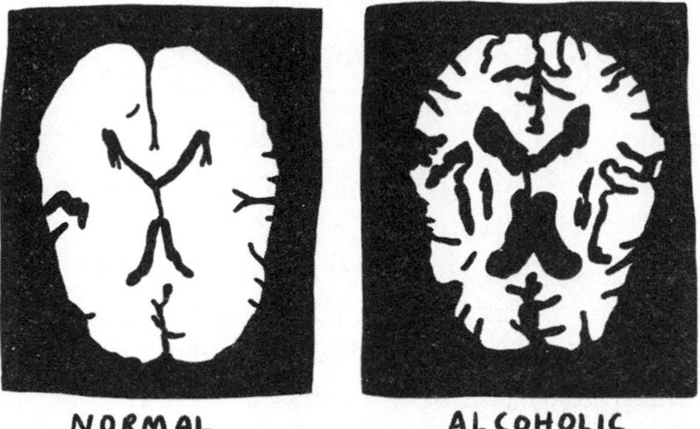

will come in the form of supportive cells and partial return of existing neurons to normal size. Once neurons themselves die, they are gone.

What is "natural" wine?

On the corner of the street in the Fort Greene neighborhood of Brooklyn where I lived in 2015 was a wine store that constantly advertised on its sidewalk chalkboard that it sold "natural wine." The next times I went to Los Angeles and San Francisco, I saw it there, too. Once I woke to the trend, it seemed that in every neighborhood where there was a Lululemon and Restoration Hardware, there was natural wine.

It felt to me like the equivalent of selling natural iPhones. Wine is the product of farming grapes; crushing and fermenting them in barrels; and bottling and corking and shipping them in opaque containers to protect them from sunlight, air, and ambient variations in temperature. Nature seems at odds with wine, if not actively working to destroy it.

I asked at the wine stores about their natural products. They consider wine to be "natural" when it contains no preservatives—specifically, none of the chemicals known as *sulfites*. They're preservatives that prevent bacterial spoilage of wine during shipping and storage. Sulfites also keep wine stable against oxidation. So, they are antioxidants—as in, the same things that people extol to justify drinking wine as a health measure.

Sulfites have been used for centuries to preserve wine. There's evidence that ancient Romans burned sulfur candles into the barrels because they realized that it helped prevent the oxidation of wine into vinegar (the process may indeed have created some disulfide compounds). Later, people began to use sulfur dioxide gas and bubble it through wine to preserve it. The sulfites used today were introduced into the winemaking process about a hundred years ago, with the synthesis of a chemical called *potassium metabisulfite.* It was adopted as a preservative for a lot of different foods, explains chemist James Kornacki, "but especially wine, where it's really, really needed." (Most wine that is labeled "organic" comes from grapes that were farmed organically, but sulfites were likely used to preserve the wine.)

Kornacki recently finished his doctoral research in epigenetics at Northwestern University. He first remembers hearing of sulfites in 2003 when his aunt announced that she was "sulfite sensitive." As he recalls, "She'd turn away wine at family parties. Back then no one knew what sulfites were. I thought it was a crazy word."

In some circles, it still is. As more and more people claim to be sulfite sensitive (reminiscent of gluten sensitivity, manifesting as an expansive array of symptoms), a similarly growing number of people are dismissive of the idea that sulfites are suddenly causing so much suffering. They say that if you can tolerate eating dried fruit—which usually contains tens to hundreds of times more sulfites than wine—then sulfites are not your problem.

Kornacki believes it's more nuanced, in that many of the sulfites in dried fruits are bound to other molecules, so they may be of less consequence than those in wine. Regardless, to him, if people believe they are sensitive to sulfites, they should have a way to get rid of them. But preserving wine with sulfites is still necessary, lest it be available only to people who live very near vineyards or natural wine stores. Kornacki set about to devise a way to extract sulfites without messing up the wine. His vision was to sell a consumer product that could remove sulfites at the last minute.

Products currently on the market to remove sulfites do so by adding small amounts of hydrogen peroxide to the wine. That reacts with sulfites, converting them into sulfuric acid. The dose is minute, but the concept is hard to market to wine purists. So Kornacki's solution is mechanical: a filter that forms covalent bonds with sulfites, sucking them up as the wine passes through.

The idea seems to speak to people. In 2015, Kornacki asked for donations on Kickstarter to bring the product to life, with a goal of $100,000, and got $157,404. I asked if he'd try it on *Shark Tank*. He said he really hates the show and would have to reconcile that with himself first. He'd also need sales numbers before he went to them. (Which I know, because I watch the show. I study the sharks. I believe we have much to learn from the sharks).

He'd have to work on his pitch, too. I asked if he thinks everyone should be taking sulfites out of their wine. "Yes and no," he said, as I mentally eliminated him from the shark tank. "If you've never had issues, that's going to be up to you. I think ultimately we should eliminate chemicals from our environment that we didn't evolve with over eons."

In that case, eliminate wine. Eliminate basically everything.

Roughly 1 percent of people have an allergy to sulfites, according to allergist Mary Tobin. This causes textbook symptoms of allergic reactions: itching, flushing, hives. Rare cases can lead to anaphylaxis. For those people, sulfites are bad. We learned this in the 1980s, when sulfites were being used on some leafy green vegetables. People had allergic reactions, and the FDA stepped in, banning sulfite preservatives on fresh produce.

The segregationist senator Strom Thurmond, also an alcohol prohibitionist, succeeded in passing a 1988 bill that he called a "warning label" on wines. No other food products, like the dried fruits that carry many more sulfites than wine, need carry the "contains sulfites" label.

The warning is similar to the "contains nuts" label on peanut butter. The wine columnist at the *Chicago Tribune* recently ranted about people returning from European vacations bloviating that they drank wine without any problems, because the wines there do not contain sulfites. He said he reminds them that wines *around the world* contain sulfites. The only difference is that wine sold in the United States must carry the "contains sulfites" label. Which suggests that these words can become self-fulfilling prophecies. It's especially easy to demonize preservatives like sulfites when a label suggests there might be reason to avoid them. It's easy to reason, "Well, I might as well avoid them. I mean, what's the harm?"

. . . Right, so **why not just play it safe and avoid preservatives?**

Preservatives like sulfites are themselves safety measures, so that's like asking, "Why not play it safe and go rock climbing without a rope?" The rope can give you blisters, and it can get snagged on an overhang and throw you off balance. There's a chance you could get strangled by the rope.

I'm sorry, that sounded dismissive. Preservatives don't seem like ropes because most of us take our constant food supply for granted. But sulfites, for example, are the reason that we are able to get bottles of wine from all over the world. The modern commercial use of sulfites has democratized wine. Some would argue that's a greater contribution to society than rock-climbing ropes.

Sulfites are also the reason that people can let wine sit and age in a cellar for years, at which point it not only doesn't sicken them, but actually gets *better*. When no sulfites are added to wine, it needs to be refrigerated, which is not environmentally ideal. It also has to be made in totally aseptic conditions, which ultimately sacrifices some flavor complexity, so the product tends to taste kind of like grape juice.

The dilemma with "natural" wine is greater than that it distracts from more urgent public health problems, or that it encourages people to waste time and money. At a practical level, encouraging manufacturers not to use preservatives means food spoils, exacerbating global food shortages.

But maybe most critical is the reminder that basing decisions around purity can easily become dangerous. To take the most extreme example, ask geneticists about the history of their field. They will probably begin to tug at their shirt collar, and they may suggest that it's "hot in here. Is it hot in here? Is that just me?" Etcetera. Because it was less than a century ago that many respected scientists advocated that poor, unintelligent, and disabled people be forcibly sterilized. This was done under the vague premise that this would purify the human species. To those who honestly feared that genes would sicken populations, this was mere precaution.

Natural wine is not eugenics, obviously, only another element on a spectrum of behaviors born of fetishizing purity. The point in such a dramatic comparison is that this is an approach to life and our bodies that can easily perpetuate dangerous and misguided beliefs.

And nowhere is it more pervasive than in the ways we talk about, or don't talk about, sex.

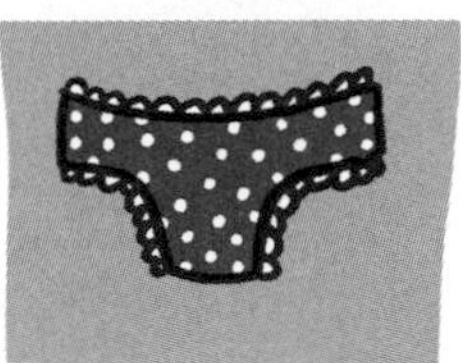
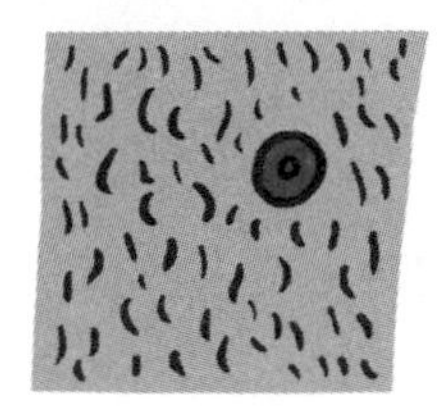
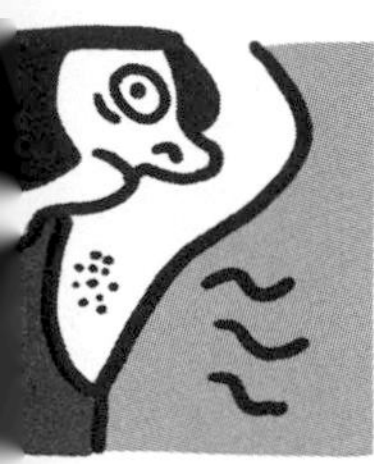

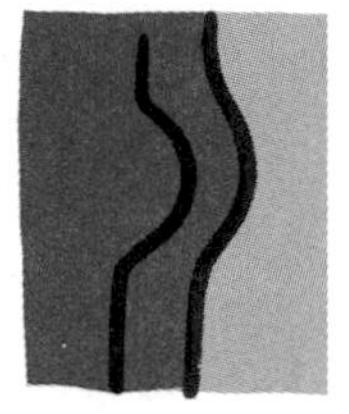

PART FIVE

RELATING

THE SEX PARTS

Holly Van Voast removed her coat on an October day in 2011, uncovering her nipples. The courtroom went silent. Judge Rita Mella of Midtown Community Court asked the bleached-blonde Van Voast, the defendant, to apologize.

Apology was a tenuous proposition for Van Voast, who was on trial for appearing topless in Grand Central Station and on the Staten Island Ferry, among other places. Or, technically, her character Harvey Van Toast was topless. Harvey looks much like Holly, but with a distinguished Dalí-esque mustache drawn in mascara. He has been many places, as a part of Van Voast's study in how people react to breasts. That day, before the judge, Van Voast's lawyer, Franklin Schwartz, urged his topless client to remember that, the legality of her immediate toplessness aside, she risked being held in contempt of court.

The day prior, eighty-nine-year-old Schwartz had been appointed to defend her by the state. He later told *The New York Times* he had never seen anything like this in his sixty-two-year career.

Van Voast did apologize and was released. She was always, eventually, released. She went on to spend most of the next two years being topless around New York, and it was often the knee-jerk reaction of law enforcement to arrest her. What began as an artistic endeavor morphed into an immersive study in both psychology and justice.

"People were constantly calling 911," Van Voast told me. "It was a firehose of opinions coming at me. You could watch it happen. It would just jam everyone's minds. I was a walking Rorschach test."

She would be arrested again in Saint Patrick's Cathedral, among other places, and brought to the police a dozen times.

Usually she could talk herself out of a cop car by explaining that baring female nipples is legal in New York. Though many women have been illegally detained, she was right: Exposed nipples have been allowable since a state Supreme Court case in 1992. *People v. Ramona Santorelli and Mary Lou Schloss* found that when the defendants were arrested for exposing "that portion of the breast which is below the top of the areola" in a Rochester public park, the statute was discriminatory. The ruling notes that the law "defines 'private or intimate parts' of a woman's but not a man's body as including a specific part of the breast. The People then have the burden of proving that there is an important government interest at stake, and that the gender classification is substantially related to that interest," which the People failed to do.

I've never run into them, but apparently a literary club called Outdoor Co-ed Topless Pulp Fiction Appreciation Society occasionally convenes down the street from my apartment, in Brooklyn's Prospect Park. They draw stares, and not only because they are reading pulps. They are part of a movement called Free the Nipple, which has been championed by celebrities like Miley Cyrus and Lena Dunham and seeks to rectify the same bias that leads people to believe that certain nipples should be illegal, fueled by policies that keep female nipples off of Facebook and Instagram. More people Googled "Free the Nipple" in 2015 than "equal pay" or "gender equality." So the idea is bigger than nipples in many ways; the nipple is the intersection of biology and sociology. Why do these structures mean so much? Why are they the dividing line between decency and nudity? Especially since everyone has them, it would seem.

Why do males have nipples?

In their 2005 *New York Times* number one bestseller *Why Do Men Have Nipples?* writer Mark Leyner and his coauthor, physician Billy Goldberg, piqued the curiosity of many. They address this question briefly, answering that it has to do with the default "female template" that we all follow as fetuses. The idea was that we all begin life as females, until in some of us the male sex chromosome "kicks in."

This was the prevailing thought for centuries—an Aristotelian idea that maleness is an active process while femaleness is a default. Stanford professor of the history of science Londa Schiebinger explained to me that we now know it's more egalitarian than that. Every step in embryonic development is an active process. The male nipple is not vestigial, and it is not a spandrel (a flourish that serves no function), but a perfect example of how fundamentally similar we are.

I learned the default-female concept in medical school, too.

"I'm sorry your medical school curriculum was based on knowledge from two thousand years ago," Schiebinger said, as noncondescendingly as possible. The same is taught at Stanford in a lot of the human biology classes. "The faculty hasn't really caught up. We are trying to spread the word, but it doesn't always get into the trenches."

Venture into the trenches of a mammography department at any hospital and you will see men coming and going throughout the day. In my experience, they often wear sunglasses. Not because they are creeps—though they may be—but because they are there to get mammograms. Men don't often talk about getting mammograms, but 1 percent of breast cancers are found in men. When men do get breast cancer, they're more likely to die of it, because they don't get checked out—because talking about breast cancer is not part of the Male Code.

That's the term used by University of Pennsylvania psychiatrist Rob Garfield. "Being a good man, traditionally, is in some ways in opposition to being a connected human being," he told me. "A successful man is emotionally restrained, keeping things close to the vest, defending your position, being in control, etcetera. If your identity is so wrapped around being that way that it prevents you from doing other things—connecting, disclosing, showing vulnerability, giving up control—you aren't developing those skills. I think guys need extra attention to this because they are working against the societally defined Male Code."

Male breast cancer may offer a case for the *least* important reason to acknowledge the fact that we all have nipples and breasts. Even the flattest-chested male has at least a few breast cells—though the odds of developing enough breast tissue to lactate is much higher among people who have ovaries that secrete the growth-stimulating hormone estrogen. The degree to which breast cells proliferate can vary a thousandfold between perfectly healthy people. Few other body parts vary so greatly. In ears, hands, ovaries, penises, spines, and other body parts, there tends not to be more than a twofold difference in size between the largest and the smallest in people.

This disparity tends to be exaggerated where it is least important, and overlooked when it is most relevant. For example, Londa Schiebinger directs a research center at Stanford devoted specifically to the effects of gender in innovation. The "default state hypothesis" has been one of her passions. Because the female was seen as a default, the pathway wasn't studied. "It's a human body. We all have the same basic plan or architecture. Then there's a cascade of genetic, hormonal, and environmental influences that move people in one direction or another."

In the first few weeks after conception, we were all devoid of any semblance of sex. We all started out the same way: one cell, and then two cells, and then a ball of cells, and then a tube of cells, and then a spine with a head, and so on.

The long-prevailing view of sex differentiation went basically like this: Inside an embryo, a few primordial germ cells migrate into the area called the genital ridges, which will become ovaries—unless the embryo has a Y chromosome. Specifically, one region on the Y chromosome known as the *SRY* gene (discovered in 1990) determines the sexual destiny of a fetus. The presence of that gene tells the cells in the gonad to become Sertoli cells, which then produce the signals that direct development of maleness—the making of testosterone, which will fuse the labia to become a scrotum, expand the clitoris into a penis, and halt breasts from fully forming. Primordial germ cells start to develop not into eggs, but into sperm.

Only in 1994 did researchers identify people who developed as "women" but had X and Y chromosomes—with intact *SRY* regions on the Y chromosome. It was later found that in these XY females, a gene called *DAX1* "can act as an anti-testis gene," actively suppressing the male pathway.

The fact that ovary formation is an active process is not a linguistic subtlety. The absence of *SRY* is not sufficient to build a functioning ovary; two X chromosomes are also required. Women with Turner syndrome—who have only one X chromosome—develop as female and begin making ovaries, but the ovaries will "fail" without a second X chromosome.

Meanwhile, XX males proved that a "functioning" (hormone-secreting) testicle can develop in the absence of germ cells. Indeed, genes can override male development even when *SRY* is present. Sex depends on the interplay of many genes, with *SRY* actively promoting the male pathway by stimulating a gene called *SOX9,* while in females proteins like B-catenin actively promote the female pathway by suppressing *SOX9*.

Nipples testify to the part of the developmental pathway that is shared, the core template from which we all arise. Nipples and rudimentary breasts form before the point at which a fetus is sexualized. By puberty, breast tissue tends to proliferate in females, but it will grow in males who begin taking estrogen. We all possess the necessary machinery, it's just operated differently by hormones.

This makes it especially odd that in many countries, women can be jailed for showing their nipples in public, while men can do so at will. American men gained this right with the passage of laws in many states in the 1930s, while female bodies are censored across the country—especially on Go Topless Day. It falls on the Sunday in August closest to Women's Equality Day (August 26, a "nationally recognized date," if not a holiday, since 1971). In the

United States, individual states have jurisdiction in matters of "public morality"—nipples among them—and so it is possible for a nursing mother to cross a state border and be arrested. I grew up a mile from the border between nipple-friendly Illinois and nipple-antagonistic Indiana, where displaying "female nipples" constitutes a class B misdemeanor. Facebook is the Indiana of the Internet.

Even where nudity laws allow nipples, women are often detained through the gaping legal loophole that is "disorderly conduct," for which Van Voast went to court. When police arrested her for indecency and realized they were wrong, as she recalls it, they'd sometimes apply other charges. Another time, they recorded it as a hygiene issue. Only in February 2013, after she had yet again been wrongly arrested, did New York City police receive a memo that they were not to arrest or detain women for "simply exposing their breasts in public."

The distinguishing factor between male and female nipples is not even the quantity of breast tissue beneath them—many men, especially when obese, have more tissue there than women. Nor can law enforcement officers see the chromosomes of a person and decide whose nipples are female and whose are male. So the female nipple is a matter of context. It is the perceived construct of femininity that, in combination with nipples, constitutes disorder.

As it becomes clearer each year to those studying embryology that each physiological outcome—including sex—requires an ever more complex cascade of gene signaling molecules and precise timing, so does it become clearer to Schiebinger that "sex is basically a continuum." One percent of the world's population is estimated to be not male or female but intersex. The estimate is rough, because many infants undergo immediate surgical or hormonal intervention to push them into the binary. The exact numbers are further confused by the paperwork that obstetricians must fill out. Germany is the only country where a birth certificate can be marked male, female, or other. "In most places, you're forced into two boxes," says Schiebinger. "The boxes may not represent reality. And why is that?"

Because we have a tendency to create order out of complexity? It's a trait especially associated with brains that are exposed to more testosterone in utero, "male brains," though similar personality effects of testosterone can be seen during gender transitioning, and changes in gray matter have been documented in as little as a month. The areas known for language processing, Broca's and Wernicke's areas, sometimes become measurably smaller. According to Andreas Hahn at the Medical University of Vienna, higher testosterone is linked to smaller vocabulary in children, and verbal fluency decreases in female-to-male people taking testosterone.

The differences also transcend hormones and innate biology, into the effects of years of (looking-glass) self-perception in a gendered world. When the recently transitioned Caitlyn Jenner explained on her reality show in 2015, "My brain is much more female than it is male," many applauded her bravery in publicly identifying with a marginalized population. Journalist Elinor Burkett contested the brain comment, among others, in an essay in *The New York Times,* writing that Jenner's claim to a female brain simply cannot be the case without having traveled through the world being treated as a woman—"having accrued certain experiences, endured certain indignities and relished certain courtesies in a culture that reacted to you as one." Femininity is not essential to biology, Burkett argues, but "a social construct that has subordinated women," and one that cannot be claimed overnight.

Meanwhile, masculinity isn't exactly doing wonders for society. In his work to help men transcend the Male Code—to develop friendships and express emotions (other than anger)—Rob Garfield has found that much of what we consider maleness is malleable. If not always easily. He draws an analogy to creatures from *Harry Potter* called Dementors. "They sort of swirl around a person and suck the life force out of them," he told me. "Social stereotypes function as Dementors in men. And one of them I named Homophobo. It's the cultural stereotype that scares men off from showing that they're interested in each other, expressing warmth, admiration, because of the cultural taboos of being seen as gay."

Men are much less likely to see doctors than women, making 134 million fewer physician visits each year. "This all comes back to the Male Code: Be strong, independent, physically tough, don't show it if you've got pain, or if you're suffering," said Garfield. "It is getting better, but it's still a huge problem."

Garfield was inspired to study male-male social dynamics after his own divorce rendered him essentially friendless in adulthood. He had the sort of friends who would help him move a couch or watch the game, but no one he could really connect with. Having conquered his own misgivings, he now runs support groups for men (called friendship groups) in Philadelphia. There men go for reprieve from expectations that they be stoic and aloof. One of the group members, a surgeon, spoke to me on condition of anonymity because he did not want people to know he was in the group. (So many layers there. Then, to top it off, he had nothing interesting to say.)

Health differences are enormous between emotionally open guys who have close friendships and those who don't, Garfield explains. The effect permeates: "Their recovery times from illnesses mental and physical, their resilience and resistance, and survival times when diagnosed with terminal illnesses, all of these things are worse among guys who don't have good social ties. They're 50 percent more likely to have a first-time heart attack. They're twice as likely to die from it."

Women have for decades been significantly more likely to die of heart disease than men—even though men are more likely to *have* heart disease. The American Heart Association attributed that to culturally driven conceptualization of heart disease as a "man's disease," which has been subsequently misdiagnosed by doctors and overlooked by women who *couldn't possibly be having a heart attack*.

To Schiebinger, this is a classic case of gender-driven agnotology. It is with an eye to avoiding similar blind spots that she works to understand sex-based bodily variation—to know where it is relevant and important to emphasize physical differences, and where it is counterproductive. She and colleagues at Stanford have attempted studies on sex and gender differences in the brain, but "could never actually do one, because it's so complicated and fraught. Talk about wars," she said, pausing. "The knee is interesting."

The knee *is* interesting. The company Zimmer now sells prosthetic knee replacements made specifically for women. Called the Zimmer Gender Solutions High-Flex Knee, the company states frankly on its site, "When it comes to knees, men and women are different."

Why would that be? It's true that archaeologists are able to determine the sex of a skeleton at a glance based on one factor: the shape of the pelvis. A female pelvis is almost always wider, ready to pass a child. That width means the femur angles slightly more inward toward the knee, creating a slightly different angle—on average—than it does in males. The Zimmer Gender Solutions knee claims to account for that angle, as well as the generally smaller female knee.

"I was initially so excited," Schiebinger recalls, upon learning of the new prosthetic knee. "Look, this knee implant for women!" She talked it up around Stanford medical school. When she took it to orthopedic experts, however, her trepidation grew. Considering the special angle of a female knee is one thing, they told her, but it's less important than a person's height. Maybe even more important is the surgeon's experience and the hospital's ratings on infection—and the patient's willingness to do physical therapy. Sometimes elucidating differences is productive; sometimes it's not.

HOW TO TELL A SKELETON'S SEX

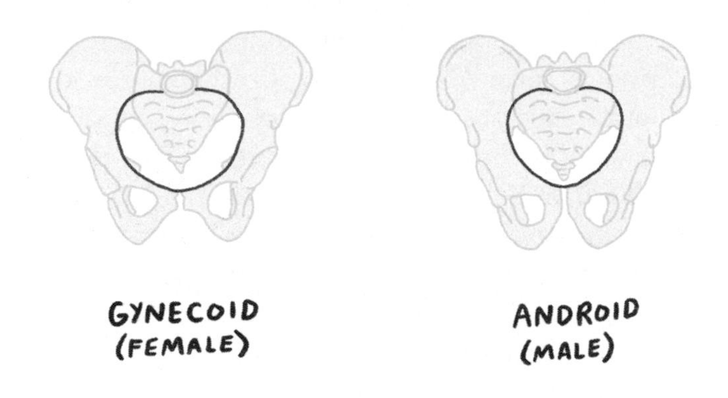

"We should be able to be different and equal," she says, ever conscious that wage gaps and other ongoing systemic gender-based discrimination are the products of ideas about sweeping, fundamental differences between men and women. The line is fine. "Very often, equality boils down to us all being the same."

It is unifying to consider that we all begin the same as sexually undifferentiated embryos, and that there is no *default* sex pathway, but rather an elaborate milieu of chemical signals yielding infinite spectra of physical characteristics that can be regarded in infinite ways.

...Then **why are nipples sexualized?**

In Poland—where female toplessness is illegal—anthropologists Agnieszka M. Zelazniewicz and Boguslaw Pawlowski have extensively studied the nature of human attraction to breasts. They write in one academic journal that "large female breasts should be perceived as attractive" to heterosexual males because they may signal an ability to bear children. Infants do not spring forth from breasts, but females with low waist-to-hip ratios and large breasts tend to have higher levels of the hormone associated with fertility. "Large breasts may also be a cue for better genetic quality," write the anthropologists.

Still, in their research it was not the case that bigger was better. To men, the most attractive bra sizes were C and D, beating out the smaller A and B, but also the still larger E. The anthropologists estimate that as breast size increases during pregnancy and lactation, "there is a possibility that breasts that are too large signal that a woman is not fertile at the moment and, therefore, less attractive, especially for short-term mates." They suggest that men may also discriminate against the largest-breasted women "due to anticipated infidelity," as other research has shown that females with large breasts are perceived as being more promiscuous and "sexually open." Females with smaller breasts are perceived not only as moral and modest but also as competent, ambitious, and intelligent. Other researchers have found that male preference for larger-breasted women can be even more deeply arbitrary, in one study coming down to whether or not men were hungry at the time that they expressed their preferences between two bust profiles. This is part of a resource-scarcity hypothesis, wherein males with tenuous access to the basic needs of daily life tend to prefer females with larger breasts.

After spending two years topless in New York City, Holly Van Voast is at once more universal and more succinct in her wisdom: "Look, everyone loves tits. They're alluring. We're taught to look for them as babies. People are programmed to love them." She uses the term *cognitive dissonance* every few sentences when telling stories about all the people who took serious and sometimes violent offense at her breasts on the subway or in the park. The worst reaction was a woman with a child who shoved Van Voast into a police barricade. "Acting like you hate them anywhere is absurd. I'm not a lesbian, but they're alluring."

When men took offense, they would often draw an analogy to their own penises. Something to the effect of "What if I went around with my dick hanging out?" Van Voast was struck by how often people drew analogies to penises. The penis-boob equivalence is cultural rather than anatomical; the anatomic analogy would be the clitoris. She came to understand the sexualization of breasts as a projection of male psychology. With sexualization of any body part comes timidity and judgment, a result of pervasive suppression of sexuality. This has tangible health effects. For example, breast reduction surgery remains more common than breast augmentation surgery, and is proven to be effective in treating and preventing back and neck pain, improving a person's quality of sleep and ability to exercise. Still, because of the stigma around all breast surgery, reduction is often not covered by insurance. To truly solve the nipple disparity, Van Voast has concluded that women need not march on Washington topless, as in the popular Free the Nipple campaign, but simply go topless around their neighborhoods in their daily lives. Though it will require fortitude. Nipple activists, she says, are "swimming to the surface of history's burdens on women to absorb amazing amounts of ridicule and abuse." But she is uniquely equipped to deal. "I have the ability to process these things, and most women don't. I don't know why. I don't know why. It's like I'm the Turing of tits."

Still, despite her powers, she gave it up in 2013. The cognitive dissonance was unbearable. She won $77,000 in a lawsuit against the NYPD for multiple wrongful arrests and moved upstate to Schenectady.

"There are huge power struggles in our society," she explained, "and the topless thing was just a way of seeing these things in a more obvious manner."

Why do penises look like penises?

Why the shaft and glans of the human penis are so much larger than the shaft and glans of the clitoris has long been a sort of unquestioned curiosity. One book that made important headway into understanding why penises have a bulbous glans at the end—instead of just being a nondescript cylinder, or even a hollow cone of the sort that is used during artificial insemination—is journalist Jesse Bering's *Why Is the Penis Shaped Like That?* In it, he describes Semen Displacement Theory.

As anyone who has spent more than a few minutes on YouTube knows, many animals have penises. Not many of them use the penis to thrust in and out as persistently or as aggressively as the human male, though. Part of this is born of tradition and the unimaginative male mind. But part of it is physiologic demand. Why should it feel good to males and females when the penis is thrust in and out? Why not, like a lion, just let it sit in there, remaining perfectly still, until it does its penis thing and deposits semen?

The reason might be the same as that for the glans's bulbous shape. Semen Displacement Theory posits that the helmet and the coronal ridge, combined with repeated thrusting, serve to pull semen out of the vaginal canal. And why do that? For the same reason males do so many things: because mating is a competitive sport. The idea is that another male might have recently deposited semen in this particular female's vagina. The job of the male and his penis, then, is twofold: to deposit his own semen, yes, but also to remove all other semen in the process.

If ever there were a physiologic argument against monogamy, it might be Semen Displacement Theory. In this way, a larger penis is advantageous for the romantic reason that it's more effective as a sort of semen shovel.

Psychologist Gordon Gallup at the University of Albany tested the mechanics of this theory. Not by having people have unprotected sex with multiple partners in succession. He used prosthetic penises. And it did seem to work.

One might wonder: Couldn't this be counterproductive, in that a penis might end up removing its own semen?

That's unlikely, because contrary to so many lyrical allusions to "making love" for periods lasting "all night long," most erections dissipate shortly after ejaculation. This fact is less popular in song, as is the fact that the post-ejaculation penis tends to become averse to further stimulation. What was moments ago pleasurable—to a degree that males would go to tremendous lengths to achieve—becomes disagreeable.

If Semen Displacement Theory holds, this would make sense from a reproductive standpoint. The male should want to keep his semen remover away from his deposit, lest his work be undone. Maybe best for him to just fall asleep.

When is ejaculation premature?

The average duration of heterosexual human intercourse is three to thirteen minutes, usually ending when the male ejaculates and becomes lethargic. Other species do not spend even this much time. Lions average less than one minute. Marmosets ejaculate within five seconds of penetration. If you talk to marmosets, they say all the better to get back to hunting and protecting their families from predatory birds.

Still, shouldn't natural selection favor the males who most quickly deposit "the goods"? (And should we ever refer to semen as "the goods"?) If you believe in Semen Displacement Theory as an explanation for the absurd shape and size of the male penis, then a lengthy thrust session may be a subconscious instinct to thoroughly scour the vaginal canal. Then, and only then, should one deposit the goods. Alan Dixson, a professor of biology at the University of Wellington in New Zealand, has indeed posited this explanation for the human predilection for prolonged sex with "patterns of deep thrusting."

Why don't males have multiple orgasms?

At the Marriott in downtown Charleston, South Carolina, scientists gathered for the 2016 meeting of the International Society for the Study of Women's Sexual Health. There researchers reported findings from the largest nationally representative study of, as they referred to it, "women's pleasure." It's not an attempt to sidestep the word *orgasm,* but to carefully note that a contraction of muscles in the pelvic floor is not the sole element of what's pleasurable about sex. Researcher Debby Herbenick of Indiana University's famed Kinsey Institute assured me that women can have pleasurable experiences without an orgasm.

It was Alfred Kinsey himself—who came to Indiana to study wasps and ended up the world's premier researcher of human sex habits—who reported in his 1953 book *Sexual Behavior in the Human Female* that most human females have the capacity for multiple orgasms. To the scientific community at the time (mostly men), this was a revelation. Virginia Johnson and William Masters at Washington University in St. Louis would go on to turn Kinsey's survey findings into hard evidence. In a lab, women stimulated with a vibrator would often have several orgasms within minutes. Some had as many as fifty.

The Study of Women's Sexual Pleasure involved three years of in-depth interviews with more than two thousand women and found that today 47 percent of women have had multiple orgasms. "Our sense is that it's possible in more women," Herbenick told me, "but it's often a partner issue. Some women become too sensitive to continue after an orgasm. Others have one orgasm and are like, 'That's sufficient.'"

After the first orgasm, she explains, most females find that different techniques build to the second. Because sensitivity is so heightened after the first orgasm, the exact same motion can be uncomfortable or even painful. It's a mistake to take that as a sign that someone isn't capable of multiple orgasms, only that they will arise in a different way. A second, third, or fortieth consecutive orgasm is usually not a result of increasing intensity, but of less direct pressure and slower movement.

While it's common for females to have this capacity, most males have a refractory period. There are, though, Herbenick tells me, "a handful of men" who "can keep having erections to the point of ejaculation one after the other." In those cases, though, the sperm count in subsequent ejaculations falls dramatically. These are not functional ejaculations, but ejaculations of pride and indulgence. If a male has the capacity to orgasm repeatedly, there would be no logic in it from a functional biology perspective. The body has almost no capacity to store more than one load of sperm at a time. Inside the body, sperm die and mutate quickly. The testicles must hang below the body because sperm can be produced only at temperatures slightly below that of the human body. Once produced, sperm can be stored only in limited quantities and live only a few days. So, why would a male have multiple orgasms if only the first one can produce enough sperm to lead to pregnancy? The real limiting factor is the tenuous life of sperm. Guys, if you want multiple, rapid-fire orgasms, evolve a better system for storing sperm. Maybe it is a sack that is sewn to the perineum? I don't know. This is why there are venture capitalists.

Females, subject to no such problem storing gametes—their eggs are present and stored away since infancy—should be free to orgasm endlessly. Yet, Herbenick and many others lament, this potential is often unrealized. The common difference between females who are sexually thriving and those who are not is less often physiological than social. Women who feel like they can talk comfortably and explicitly with their partner are far more often satisfied. While multiple orgasms are great, Herbenick's urgent point is that when unsatisfying sex becomes the norm, it warrants discussion. The study also found that women who felt that they could talk with their partners specifically about what makes sex pleasurable for them are eight times—*eight* times—more likely to be happier in their relationships. Again: eight times. A culture where talking about sexual pleasure is taboo cannot earnestly espouse "family values."

As Herbenick put it, "A major family value would be talking explicitly about sexual pleasure."

How do you responsibly inform a clingy ex-lover that you have been diagnosed with gonorrhea (by phone)?

"Hi, Derek."

"Hey! Oh my gosh, it's so good to hear from you! I've been texting you, like, nonstop. Did you get my texts? I texted you so much."

"Sorry, I'm in a tunnel, so I might lose you. I just wanted to say that I've been diagnosed with gonorrhea, and it's a drug-resistant kind, and you need to get tested. Will you get tested?"

"Super gonorrhea?"

"Yes, that's it. Will you get tested?"

"Yes, sure, do you—"

"Sorry, ah, tunnel!!!" [Hang up]

Only in some places is it legally required that a person notify all recent sexual contacts when diagnosed with a sexually transmitted infection. It is always a responsible practice, even if you would get some vindictive pleasure out of learning that a particular contact was ill. Because before you know it, there's a syphilis outbreak that you could have prevented with a single call/text/email/Edible Arrangement. As an easy alternative to directly contacting a person with whom you no longer wish to speak of sex, there are anonymous notification services that you can use, and they are free, like the one at Dontspreadit.com. Don't use these services to prank people.

How big is the average clitoris?

In the years that Londa Schiebinger has been teaching a class on sexuality at Stanford, she has noted that "most" of her male students "can tell you the length and diameter of their penis,

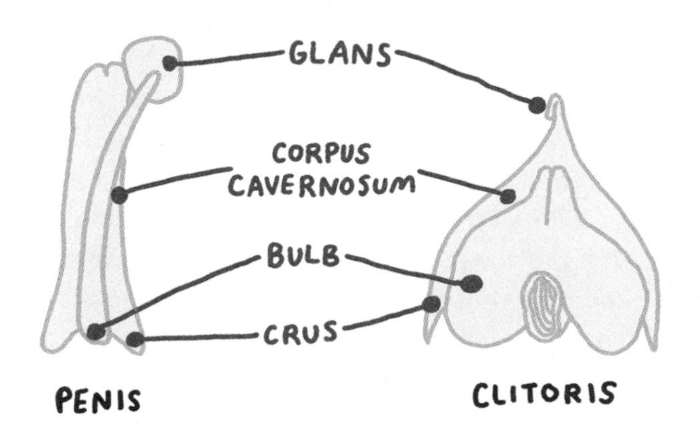

both flaccid and erect." Meanwhile, her female students tend to have "no idea" how big their clitoris is, nor of the variation in size of the clitoris among women. When the magazine *Science* published an article titled "How Big Is the Average Penis?" in 2015, it remained among the most popular articles on their website for months.

(The answer, based on a review of seventeen studies of 15,521 males worldwide, was an erect 5.1 inches [13.1 cm] long and 4.6 inches [11.7 cm] around. Not that it matters. Unless it does.)

Less known is that the average clitoris is not so far from that size. That fact is less important to Schiebinger than the fact that most people don't know it.

In the sixteenth century, when the world was busy building phalluses in every major city, the famed French anatomist Ambroise Paré would reference the clitoris only as the "obscene part." His contemporary, Italian surgeon Realdo Colombo, claimed to have discovered the clitoris, in much the same way that European traders discovered the long-inhabited American continent. Colombo called the clitoris "the seat of women's delight," likening it to a tiny version of the penis. That perspective would be propagated for centuries and is still found in embryology lessons today.

Schiebinger considers this a perfect example of the way a culture can inadvertently create ignorance. Both males and females know more about male than female anatomy. Societies keep it so. Clitoral agnotology.

It was not until 1971, during the women's liberation movement, that a group of women in Boston compiled the reference book *Our Bodies, Ourselves* to serve as "a model for women who want to learn about themselves, communicate their findings with doctors, and challenge the medical establishment to change and improve the care that women receive." The book is still in print in many languages, spreading worldwide the knowledge that the clitoris is more than a prehensile penis (as had been reported by the researchers Virginia Johnson and William Masters just a few years earlier), but a distinct organ composed of the glans (the part described by Colombo, akin to the head of the penis) as well as the much larger shaft and crura, extending beneath the vulvar skin. So it is not possible to measure one's own clitoris, and to know a true average clitoris size would require access to many corpses. Even then it would be difficult, because the clitoris is largely spongy tissue that engorges with blood when aroused, like the penis, so it is best studied in arousable bodies.

Recently, MRI studies have allowed researchers to approximate the volume of a nonaroused clitoris as ranging from 1.5 to 5.5 milliliters. When a person is aroused, the clitoris roughly doubles in volume, increasing pressure on the nerve-dense area at the anterior vaginal wall. The glans alone averages 0.09 to 0.17 inches (2.4 to 4.4 mm) wide and 0.15 to 0.26 inches (3.7 to 6.5 mm) long. Females on the smaller end tend to have fewer orgasms. No such relationship exists in men. Yet no article on clitoral size has seemed to pique the interest of the readers of *Science.*

Schiebinger's point is not that size matters, but that clitoral ignorance is built into culture, held strongly in contrast to penile talk and homage, which is ubiquitous. (When people do talk about female sex organs, they tend to refer to the *vagina,* which is a whole different thing.) This discrepancy in attitudes between organs—that are in many ways the same—accounts for and is explained by many societal ills.

Does the G-spot exist?

The concept was named in 1981 for the German gynecologist Ernst Gräfenberg (a man), who described it three decades prior while looking at the role of the urethra in sexual stimulation. He described it as an "erotic zone" on the anterior wall of the vagina "along the course of the urethra."

Northwestern University gynecologist Lauren Streicher told me that most in her field today believe that the spot exists in most females. To deny its existence is to open oneself to criticism of denying female sexual pleasure and liberation. No one was debating the presence of male balls. (Some people believed in their presence so strongly as to hang them from the backs of their trucks.) Still, unlike male balls, no autopsy or medical imaging study has ever definitively identified a structure that clearly corresponds with the G-spot. Tissue known as the "urethral sponge" does sit in the vicinity and fill with blood during sexual stimulation, just as the clitoris and penis do, giving it a distinctly ribbed texture, like a bike tire. (Never liken it to a bike tire.) Some believe that this engorgement is not a part of sexual arousal directly, but a simple matter of cinching down the urethra to prevent it from emptying the bladder during sex—a practice not everyone is into.

The same happens in the male penis. This is why most people can't urinate while aroused. It's also why people get "morning wood." While the voluntary muscles of the body are relaxed during sleep, filling the penis or urethral sponge with blood can be a last-ditch effort to prevent micturition, the clinical term for urinating. If ever you must talk of this over tea, use the term "micturition."

Just as engorgement of the sex organs should not be mistaken for love, or even sexual arousal, the existence of an "erotic zone" does not constitute a distinct sex organ. Londa Schiebinger notes that another "G-spot" is located between the perineum and rectum. It's a nerve-rich patch of erectile tissue sometimes referred to as the perineal sponge, and it's accessible along the posterior vaginal wall and through the rectum (potentially accounting for orgasms during anal sex).

Even the experts who decry the concept of the G-spot concede that there is something important about the area. G-spot denier Susan Oakley, a urogynecologist in Kentucky, contends that it is not an entity unto itself, but an extension of a mechanism to stimulate the clitoris. She calls this the C-spot. Still others carefully note that the G-spot concept is an oversimplification that can lead people to focus too heavily on one particular area.

These ideas coalesce in an exhaustive review of G-spot-related research led by endocrinologist Emmanuele Jannini at the University of Rome, which explains that the anatomy of the pelvis is dynamic, and posits that the "legendary" G-spot is best thought of not as a structure, but as a set of interactions among the clitoris, urethra, and anterior vaginal wall. This they call the clitourethrovaginal (CUV) complex, "a variable, multifaceted morphofunctional area that, when properly stimulated during penetration, could induce orgasmic responses."

CUV complex is more nuanced than *G-spot*. It's also less named-for-a-man. Though, to be fair to Gräfenberg, he did invent the first IUD, using the cultural advantages offered to him by his testicles to advance women's sexual independence. Most of that credit usually goes to birth control pills, but his device is still today the more reliable and affordable measure. And for his work, he received not wealth and acclaim, but imprisonment by the Third Reich. After four years, he was rescued by Margaret Sanger, who would go on to found Planned Parenthood, and who negotiated his release via Siberia.

Gräfenberg came to work for Sanger's "research bureau" in New York (also known as the nation's first birth control clinic) and to volunteer for Alfred Kinsey's first studies. If ever maleness were excusable in an eponym labeling the female body, the case could be made to let Gräfenberg keep his spot. It has other merits; it's probably less likely to induce word aversion than is *perineal sponge* (certainly not as much aversion as the Word). And paramount in these matters of "mystery" and silence is that people overcome aversions. This happens by listening and talking.

Why isn't there a "female Viagra"?

In the summer of 2015, a petition garnered more than sixty thousand signatures from supporters of a drug being touted as "female Viagra." The name was flibanserin, and it had been rejected by the FDA twice because it had not been proven safe or effective. The signers of the petition, part of a movement called Even the Score, supported both by some women's rights organizations and by Sprout Pharmaceuticals, suggested that more invidious factors were at play.

"Women have waited long enough," the petition read. "In 2015, gender equality should be the standard when it comes to access to treatments for sexual dysfunction."

Vital as the spirit of the protest was, it ignored that Viagra has the same effect on the clitoris as it does on the penis. And that engorging these organs with blood is a narrow-minded approach to "sexual dysfunction."

Look up erection physiology on WebMD, and you are given information exclusively about penises. It is not inaccurate, but it is incomplete. Penises are glorified sponges encasing a tiny urethra through which the goods travel. You can trust WebMD when it says that penile erections happen when blood fills the spongy tissue called the corpus cavernosa ("cavernous body") that runs along the shaft.

When blood flow is impaired, as when a person gets atherosclerosis in the vessels that supply the penis—the same process that happens in the heart, killing more people than any other disease—the penis cannot become erect. Viagra (sildenafil) was discovered while researchers were looking for a medication that could lower blood pressure by dilating vessels. It didn't work well as a drug for cardiovascular disease, but it did successfully fill the corpus cavernosa, engorging the penises of research subjects. So was born one of the most commercially successful drugs in history. In its first year on the market, sales were over $1.2 billion.

Of course, because the mysteries of the clitoris are not mysteries, we know that females have corpus cavernosa as well. Females get erections in the same way as males. Because most of the clitoris is internal, it requires advanced imaging physics to study female erection. Recent MRI studies of women viewing "erotic videos" showed an average increase in clitoral volume of 90 percent. One additional layer of tissue makes penile erections more rigid than clitoral erections, but the process is effectively universal.

I know women who swear by Viagra recreationally, though none who have obtained it legally. (Which is dangerous; it's a serious drug with profound cardiovascular effects that have killed people.) That Viagra can enhance the female sexual experience was borne out in clinical trials as far back as 2003, when researchers at UCLA's department of urology found that, as expected, the drug increased what researcher Laura Berman called "sensations of warmth, tingling and fullness." The quest for a "female Viagra" distracts from the idea that erections are a universal element of human sexuality.

Where Viagra fails is, like so many pills, conceptually. Being turned on is the product of a milieu of elements psychological, neuronal, hormonal, and vascular. Viagra affects only the last of these. It does not affect the mind, making a person want to have

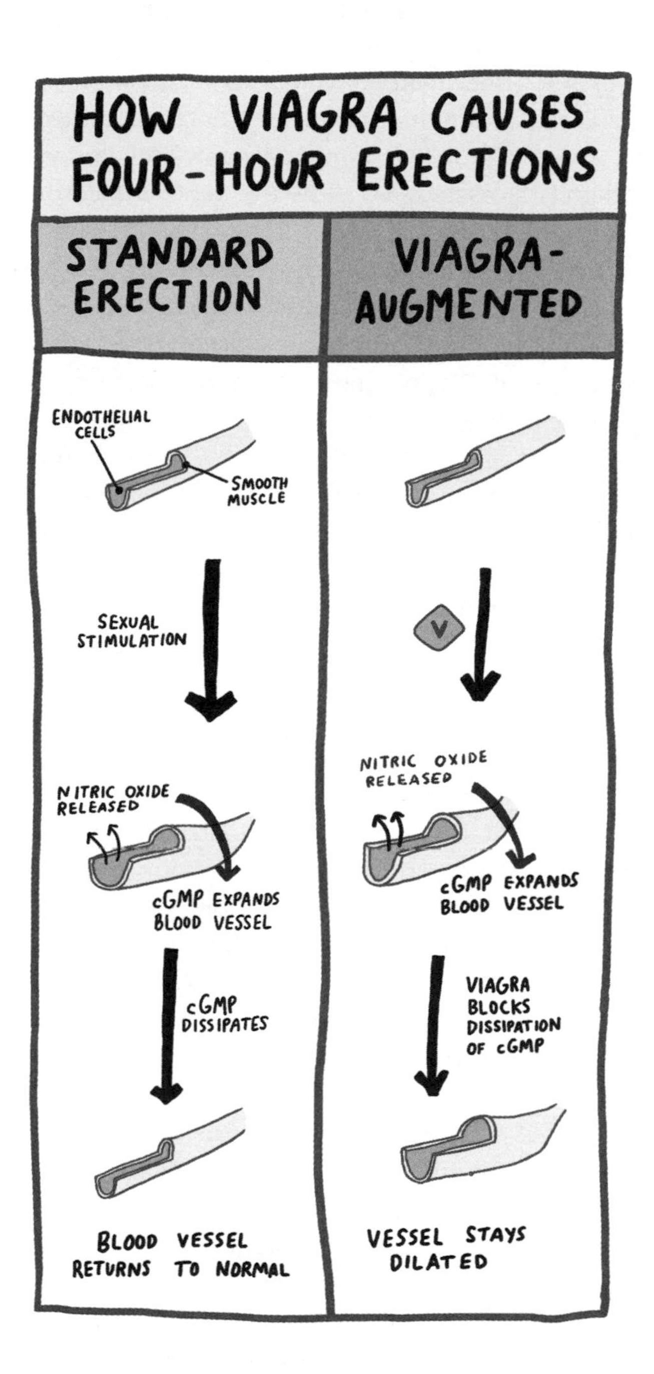
HOW VIAGRA CAUSES FOUR-HOUR ERECTIONS
STANDARD ERECTION
VIAGRA-AUGMENTED
ENDOTHELIAL CELLS
SMOOTH MUSCLE
SEXUAL STIMULATION
V
NITRIC OXIDE RELEASED
cGMP EXPANDS BLOOD VESSEL
NITRIC OXIDE RELEASED
cGMP EXPANDS BLOOD VESSEL
cGMP DISSIPATES
VIAGRA BLOCKS DISSIPATION OF cGMP
BLOOD VESSEL RETURNS TO NORMAL
VESSEL STAYS DILATED

sex. It does not affect testosterone, making a person more masculine (as Cialis commercials imply). It certainly does not make the penis larger (as Cialis commercials certainly imply). It serves only to dilate the vessels that allow the penis and clitoris to fill. It decouples the mind and body.

The drug flibanserin that was heralded in the news as "female Viagra" in August 2015 after its approval by the FDA is an entirely different pharmaceutical approach. Addyi, its brand name, is a product of Sprout Pharmaceuticals, and it acts exclusively in the brain. It is classified as a "multifunctional serotonin agonist antagonist," meaning that it affects levels of the neurotransmitter serotonin in many ways that are not fully understood. Serotonin is, of course, the same neurotransmitter targeted by many medications that are sold as antidepressant, antianxiety, and antipsychosis. Even Sprout Pharmaceuticals does not know how Addyi works, but they did accumulate enough cases to convince the FDA—on the third attempt—that some women who had been given the label "generalized hypoactive sexual desire disorder" did see improvements in libido.

In a study conducted by the pharmaceutical company, Addyi showed no benefit to daily sexual health. The company then convinced the FDA to consider the effect on a monthly basis, and it did appear to increase the number of "satisfying sexual events"—by 0.5 to 1 event per month. That was enough to qualify the drug as "effective." People who take Addyi are warned not to drink any alcohol at all, as those who do are given to passing out. (Oddly, the company tested the alcohol interaction only in men.)

Spontaneous loss of consciousness aside, the evidence that women will be better off for the existence of this pink pill is "very weak," as diplomatically put by Judy Norsigian, cofounder of Our Bodies Ourselves, and Diana Zuckerman, president of the National Center for Health Research. They called the Even the Score petition a façade to circumvent proper drug safety testing under the guise of feminism. The National Women's Health Network and Jacobs Institute of Women's Health spoke out against the FDA approval as well.

At the same time, Pfizer publically abandoned the idea of marketing Viagra to women in 2004 because, as *The New York Times* put it, women "are a lot more complicated than men." Filling their organs with blood does not solve the problem of low libido, which is more common in women than it is in men. This problem is much more complex than a simple lack of blood flow, and to overlook its causes in favor of simply diverting blood to the genitals would be dangerous. To that end, the truest "female Viagra" is cultural prioritization of female sexual health, and that won't come in a pill.

Can I use hand sanitizer as deodorant?

Yes. This is based on an experiment where I was the lead researcher. The sample size was one person. I did it a couple times at the beginning of the Great Soap Abstention of 2016. Before my body was "used to" going without soap and deodorant, there were days when I needed urgent intervention on the go. Hand sanitizer kills the bacteria that produce odor, plus it gives you that nice alcohol smell. The problem is that it also kills the bacteria that do not produce odor. Whatever it does in collateral damage to the *armpit microbiome* seems to, in my experience, result in the prompt return of odor-causing bacteria.

How dangerous are tight pants?

From an epicenter in Australia in the summer of 2015, fear of skinny jeans ricocheted around the world through major media outlets in twenty-four hours. After a busy day of helping a friend move boxes, a thirty-five-year-old wearer of tight pants lost feeling in her feet. And then they became limp. As people who can't move their feet tend to do, this one fell over. She lay on the ground for several hours, stranded, unable to get up, the Associated Press reported. She landed in Royal Adelaide Hospital, where doctors had to surgically remove her pants.

"We were surprised that this patient had such severe damage to her nerves and muscles," the attending physician Thomas Kimber said at the time. Kimber and colleagues ruled that as a result of prolonged compression, not only had the woman's nerves become unable to communicate with one another—an extreme version of what happens when one's arm falls asleep—but she also had some breakdown of the muscle cells in her legs, through a process known as rhabdomyolysis, which is also common in extreme athletes.

(The doctors made no mention of the additional problem that while in skinny jeans it's difficult to put a phone in your pocket. Which might come in handy if you find yourself stranded on the ground, or elsewhere.)

After four days in the hospital and a few more days of numbness, the woman recovered fully. Her case remains an isolated oddity. Dig through medical journals and you can find a story where almost any garment has injured a person. There was even a case of the reverse, during the first epoch of skinny jeans, in 1983, when they appear to have saved a man's life.

The twenty-two-year-old's pelvis was crushed in a car accident, and he was rushed to London's Westminster Hospital. He was wearing what doctors would later describe in the *BMJ* as "tight fitting jeans with a 7.5 cm [wide] belt." Despite tremendous destruction of the bones of his pelvis, the man was awake and conversant with the medical team, who deemed him to be in "stable condition" for twenty-five minutes. That is, until they removed his jeans. The man's pulse immediately disappeared.

Surgeons rushed him into the operatory and opened his pelvis, where they found huge blood clots and major vessels that continued to gush. Usually it is injury to these arteries (the iliacs) that leads a victim of pelvic injury to die of what will be referred to colloquially as simply "internal bleeding." People similarly crushed in car crashes often don't make it to the hospital. But in this case, just as anyone is taught to apply pressure to stop a wound from bleeding, the tight jeans had been controlling the loss of blood and helping to form a clot.

The surgeons were able to stem the bleeding, and eventually the man recovered. They went on to warn their colleagues in a

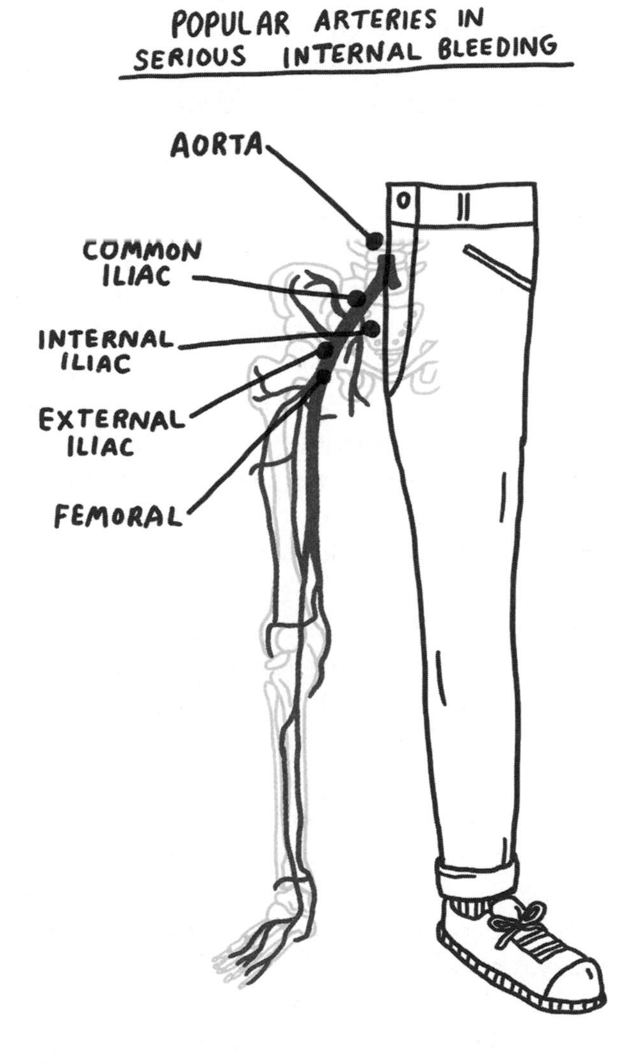

report that while it is standard practice to cut the clothing off trauma victims immediately upon their arrival to an emergency department, that might not always be wise. Indeed, some military units deploy "anti-shock trousers" that inflate to hold pressure throughout a person's lower half after a severe injury. This way, even if the soldier is going to lose his or her legs, there will be enough blood left to keep the brain supplied. As in the case of the young British man, the surgeons wrote, "A similar function may be performed by tight-fitting garments, and their role in the management of severe injuries may be very important."

Promising as therapeutic tight pants may be, their widest-reaching health consequence may be psychological. According to a press release from the department of plastic surgery at New York University, the trend in tight pants has caused a dramatic uptick in labia reduction surgery, or labiaplasty.

I know, I wasn't expecting me to say that either. But the subject of the email I received from a publicist for NYU's Langone Medical Center was pretty clear: "Tight Pants Are Causing a Plastic Surgery Trend." A quick click and, yes, I learned of an increase in labiaplasties.

Between 2013 and 2014 alone, according to numbers from the American Society for Aesthetic Plastic Surgery, the United States saw a *49 percent* increase in labiaplasties. Because of . . . pants, though? The publicist encouraged me to get the full story from Alexes Hazen, a plastic surgeon at NYU who has become a de facto expert in labial resizing and contouring. So I called her. And according to Hazen, even though "it's the kind of thing that sounds really painful"—because the density of sensory nerves in the genitals is higher than almost anywhere else in the body—people actually recover quickly.

"But, no, it's not an issue of pants," she told me. Which was not what I was expecting, given that it was supposedly the thing she had an interest in talking to journalists about, according to the publicist. "I don't think it's the tight pants," she continued. "People have been exercising in spandex and wearing tight jeans for a long time. I think the issue is pubic hair."

And so we were now discussing pubic hair. As a cosmetic surgeon with expertise in the pelvis, Hazen ends up with her finger on the aesthetic pulse. She found no reason to believe that the spike in procedures was a response to an epidemic of spontaneous labial growth (because that's not a thing). Rather, her busy practice hinges on the fact that "fifteen or twenty years ago, women had pubic hair. Now young women don't have pubic hair, pretty much."

She admits that's overstatement. But it's a seemingly clear case of technology leading medical care. Specifically, two technologies that brought the labia into the spotlight. The first is laser hair removal, which in the pubic region is now "very, very common," she says. That has elevated the status of labia from something relatively unseen to something that is now quite visible.

The other technology is the Internet, which—many argue—facilitates access to pornography. Hazen cannot overstate the role of pornography in the labiaplasty trend. The industry has created and long perpetuated an aesthetic ideal that is very specific. Now, though—because, as Hazen puts it, "you can Google something about a cat and end up on a porn site"—it's difficult not to let that specific aesthetic ideal creep into one's sense of genital peace.

As in most every story of a surgical trend that we term "cosmetic," it is important to note that this procedure did not begin with that label. As Hazen said, "There's always going to be a percentage of people who have—I wouldn't call it a deformity, but just excessively large labia—that could interfere with comfort in certain activities like bicycling, in certain clothing, like yoga pants."

The increase in procedures in recent years has been almost solely among people with cosmetic motives, but labiaplasty has been performed for decades to alleviate physical pain in a much smaller subset of people. At NYU, about 10 percent of patients who undergo labiaplastic surgery do so because of pain and discomfort. Hazen is, like most plastic surgeons I spoke to, quick to add that she is not implying that physical pain is more valid than existential angst. A primarily functional reason is not necessarily more valid than a primarily cosmetic concern. As the surgeon put it, "Appearance can also cause discomfort, emotionally."

Labiaplasties for that purpose do seem to be effective. One study found that just three months after the surgery, 91 percent of patients showed improvement on what's known as the Genital Appearance Satisfaction scale. This scale relies on self-report. There was no placebo group—the subjects all knew they had the procedure, and they knew they were supposed to feel better about their labia. Knowing what you're supposed to feel—and *wanting to feel* the way you're supposed to feel—is a powerful force.

But the rise of labial surgery is not just a story of technology leading psychology and becoming medicine. It's also a story of advertising leading ignorance. According to Nielsen, NYU Langone Medical Center spent $22 million on advertising in 2014. NYU is leading the country in a surge of direct-to-consumer advertising by major medical centers. The practice, once taboo in the profession, a sign of quackery, is now commonplace and rapidly increasing.

That advertising money is spent not just on billboards and commercials, but on attempts to influence journalists. The press release I received about tight pants was sent to me by a publicist who is paid by NYU to spread the word about labiaplasties. The goal was not necessarily to get me to write an adoring piece extolling the amazing advances in labiaplasties, but at least to plant a seed in the public mind. This is one of many seeds that normalize a practice. Eventually people get over the idea that it might be weird to have this surgery.

Eventually, well, *maybe I need one?*

The systemic scourge of yoga pants and genital dissatisfaction may well have produced a story that interests people—one they'd click on. It's the same psychology that led to global news outlets warning the world of the dangers of skinny jeans in the wake of the Australian woman's story. The cautionary tale here is not one of pants, but of media.

I went back to the publicist and asked where she got the idea that the surge in labiaplasties is due to tight pants. If not from the primary labiaplasty surgeon at NYU, then from whom? Was this based on any kind of research?

"I will try to send you more statistics about the connection between yoga pants and labiaplasty," she replied. "But that may be difficult to measure precisely."

I went back and looked at the prior emails, and I wondered if maybe I had dreamed the whole thing. If maybe my pants were too tight, or not tight enough.

What can I do to help my children understand their bodies and sex in a positive way? I think that's called sex-positive parenting. I know kids are often taught that sex is a bad thing and/or they learn that words like "penis" or "vagina" get a reaction and use them for attention. How can I teach boundaries without making children afraid of their bodies? I think it's a matter of making it clear what is okay in private and what is okay in public, instead of scolding and saying stop when little kids touch themselves. You don't have to answer this, because I just did.

I'm into the idea of sex-positive parenting. I'm also into sax-positive parenting, where when a small child is playing the saxophone in public, you actually encourage it rather than swatting it out of his hand.

How does ectopic pregnancy cause shoulder pain?

NORMAL PREGNANCY

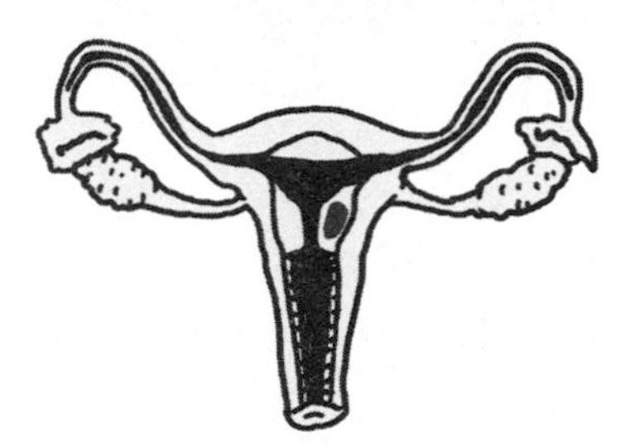

ECTOPIC PREGNANCY

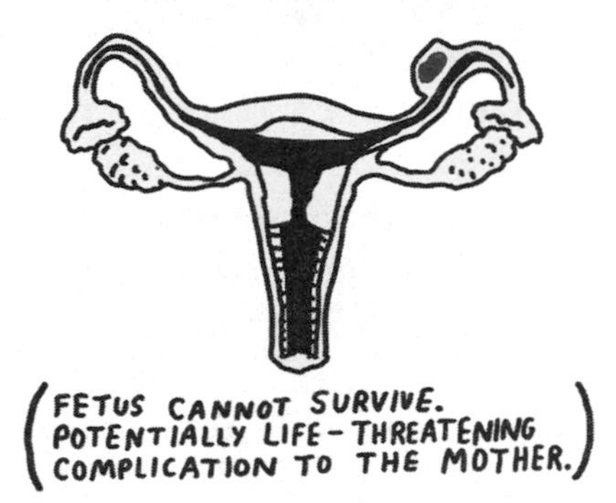

Ectopic pregnancy technically means the pregnancy is someplace outside of the uterus. The egg is supposed to float down from the ovary, through the Fallopian tube, and into the uterus. If it gets stuck in the tube along the way and fertilized, it can implant in the wall of the tube.

That is never a pregnancy that can lead to a live birth, and it is often a threat to the life of the mother. Once the ectopic pregnancy ruptures, blood pours into the abdominal cavity. Most of the visceral organs don't sense pain, but a nerve that supplies sensation to the shoulder also runs along the diaphragm, where it can be irritated by the blood. This happens around two hundred thousand times every year in the United States alone. Sometimes shoulder pain is the first sign of this medical emergency.

Are doctors trained in gender transitioning?

Lyle "Cac" Cook has a gray ponytail that might have been stolen from Willie Nelson. Though Cook doesn't seem like the stealing type. His colleagues introduced him to me as Dr. Cook, but the first thing he said upon shaking my hand was a smiling "I'm not a doctor."

By 2014, Cook had made a name for himself at his clinic in the small community of Chico, California. It happened quickly. He is a physician assistant (PA) who had developed something of a specialty in caring for transgender patients. In many places where primary care physician shortages are worst, PAs are taking expansive roles in seeing patients in near-autonomous capacities.

Caring for trans patients specifically wasn't something he set out to do. Just a couple years prior, a first-time patient had come into the clinic seeking care for her transgender daughter. Cook had no idea what to do. When he finished graduate school in 2007, he'd had no exposure to transgender health in the classroom or on his hospital rotations. (It was not an area in which I needed to demonstrate competency in order to obtain my MD in 2009, either, or to obtain medical licensure. Although some schools are beginning to incorporate it into medical school curricula.) Motivated to help a woman who had come to him with her daughter, born male, Cook read through the materials issued by the World Association of Transgender Health. Utterly unprepared for this scenario, he drove south to San Francisco and shadowed a doctor who had been treating patients to learn what to do for—even what to say to—his patient.

"It's been there all along, but there's been no word for it," Cook told me, reflecting in a workroom in the back of Transgender Health Program at St. John's clinic in South Los Angeles, where he now practices. In a matter of years, he has seen more transgender patients than most any health care provider in the country. "The similarity in the stories I hear, especially among older patients, is that they didn't have a word for how they felt. They felt like they were normal. It wasn't until someone pointed out that they were abnormal that they were taken aback and made to realize, 'I'm not right,' or 'There's something wrong with me.' From the time you're born, you're told you're a boy or a girl. Those are the first words out of someone's mouth when the baby comes out."

A determined Cook started training informally with a team in Sacramento that was treating transgender patients. On Friday evenings after his clinic shift in Chico, he would drive ninety minutes to the state capital to see patients with that team.

Eventually, he felt comfortable offering estrogen treatments to the young woman. And soon after he did, there were five, and then ten, and then twenty transgender people seeking his care. Even in the small community of Chico—home to eighty thousand—he soon had fifty trans patients who were coming to see him at the urgent care clinic. Before that, they had either been untreated or been regularly going to Sacramento or San Francisco, three and a half hours away. Cook took the swell of interest as a sign of just how needed this service was.

Then one of his patients moved to Los Angeles. He phoned back to Chico, distraught that he couldn't find a medical provider. "I really didn't want to come to L.A., such a huge town," Cook told me. "But I researched it a little bit and realized there was a huge need down here. My wife and I talked about it and said this is something we have to do."

Cook now spends all of his time caring for transgender patients at St. John's. He provides hormones, referrals for surgeries, and treatment for general health problems. On the day I shadowed him in September 2015, for example, he had to inform one patient that she had diabetes.

Aside from Cook, it was important to St. John's president and CEO Jim Mangia that his staff be entirely transgender. The clinic is run by Diana Feliz Olivia, whose story speaks to why that shared identity is so consequential.

After finishing her master's in social work at Columbia, Olivia became the trans program coordinator for the Hispanic AIDS Foundation in Queens. She did that for a while, but because she lived in Harlem, the commute to Queens was unbearable. And she had always wanted to live and work in Manhattan. Having grown up in Fresno, she wanted "the full Big Apple Experience." (As she puts it, "I wanted to be Sarah Jessica Parker a little bit.")

A job came open at Housing Works, the largest HIV advocacy organization in New York City, and they were looking for a transgender program coordinator. So Olivia got the full Big Apple Experience for two years. But then she got an unexpected call from her mother in Fresno. Olivia vividly recalls her mother saying she was "at a point where she was ready for me to begin my relationship with her." When Olivia had come out as trans in 2003, her mother was far from accepting. "I needed to leave California," Olivia recalls, "for her to deal and process losing a son and gaining a daughter."

When her mom invited her back to California, Olivia gave up the Manhattan dream and returned to be with her. She started working with a federally qualified health center that provided care for rural farm workers. Three years later, familial wounds mended, Olivia was asked to lead the St. John's Transgender Health Program. She knew it was a once-in-a-lifetime chance. She moved to L.A. to lead the team.

She believes that her identity as a trans first-generation Latina is critical to the success of the clinic. "Well, now there's someone who looks like me, talks like me, acts like me," she posited in the voice of the hypothetical patients who might initially have been afraid to come to the clinic. "When you have providers and leaders who look like the patients, they feel more welcome, more safe."

Can I get syphilis from oral sex?

I'll call her Claire because, as a patient at the transgender clinic, she requested anonymity. She had contagious buoyancy, even when she was sitting on a table covered in tissue paper looking up at Cook and me in a fluorescent-lit room named and designed to make her the subject of examination. She excused her appearance as having just rolled out of bed with an offhand joke. "I'm the bearded woman this morning."

A few weeks ago, Claire's mouth turned white. She diagnosed herself with oral thrush, which is by far the most common cause of acute whitening of the mouth. The cause is overgrowth of the fungus *Candida albicans,* which spreads throughout a person's oral cavity when the normal oral microbiome is compromised. Fungal plaques paint everything white.

The moist, dark membranes of a person's mouth are actually a lovely habitat for fungus. The two reasons it doesn't grow constantly—that our mouths aren't like word-emitting wheels of blue cheese—is that the billions of bacteria compete with the fungus for nutrients, and the rest of our immune systems also push back against candida overgrowth. But when the immune system is impaired, or when a person's normal bacteria are depleted or unwell—as often happens after taking antibiotics—the candida can "win." They celebrate by creating white biofilms that stick to the tongue and cheeks. The biofilms suck up water and create dry mouth, and then start working their way down a person's esophagus.

Before that could happen, Claire came to St. John's. Cook prescribed some oral antifungal rinses. Claire also went on the still popular "anti-candida diet," which is one of the seductive disease-fighting diets peddled by the Internet at large, despite thorough scientific de-pantsing over the years. It is predicated on the idea that the fungus needs sugar in order to survive. So a person has to spend days eating only raw vegetables (and oils) and, for some unclear reason, avoiding caffeine, in order to "starve the Candida." It's like nutritional chemotherapy, pushing a person to the limits of hunger and joylessness, until the fungus dies but the person does not.

Which is sad because cheap and effective antifungal treatments exist. Affordable rinses ask only that a person swish for one minute twice per day, for a few days. During this time, a person can continue to enjoy a calorically adequate, balanced diet. The person may also drink coffee, and the constituents of the antifungal rinse are no more "unnatural" than having a mouth full of fungus.

After the rinse, Claire's oral cavity returned to pink, the ecosystem of microbes therein returning to balance. But as is the case with so many stories of oral thrush, this was not really a story about oral thrush. The fungus is usually a bellwether. She then developed a rash, which was rare in that it included the palms of her hands and the soles of her feet. Two days before I met her, she had come to the clinic to show Cook. He thoughtfully ran some blood tests. She had been having some "really, really bad" muscle aches, but was otherwise okay. Cook tested her for Lyme disease and syphilis. Then he called her up and told her the result, and that she needed to come back in.

"I tested positive for syphilis," she explained, unnerved. "Which was crazy because I'm not a sexually active person." Her last sexual encounter had been about six months earlier. "I did fool around lightly with a guy I was talking to. It was in no way full-penetration sex at all," she went on. "Full disclosure, speaking medically, I had him in my mouth for all of thirty seconds. We had been drinking. I had been seeing him for four weeks. I did research on syphilis, and apparently that's enough. If he had any pre-ejaculate. And you know what, he just ghosted me. After that night."

Syphilis can be transmitted through oral sex. The vast majority of syphilis cases are transmitted between males, so it's a disease where a person's anatomical sex can inform a doctor's diagnostic process. (In the same way, the electronic medical record system often erroneously prompts Cook to order pap smears for every patient who is entered in the system as female.)

The rash occurs in the second stage of syphilis, which Claire knew. Typical of the modern medical patient, she was an overnight expert on her particular diagnosis. "It's really good that we caught it now," she explained to me.

Syphilis is a spirochete bacterium that has been with humans for as long as recorded history. Still, more than fifty thousand Americans alone get it every year. And the number of syphilis cases in the United States has almost doubled over the past decade.

"And all of the symptoms from the first stage, I now realize, happened. Remember hair loss?" she looked to Cook. "I had no concept that this could even be a thing. I thought it was stress."

Which it also could have been. She initially almost refused the test, because this couldn't possibly be "syphilis—that disgusting word that I curse." Now she describes herself as "beyond grateful." She asked Cook to apologize to the rest of the clinic staff for having scoffed at the initial suggestion that her symptoms might be syphilis. "Because I could've been in really deep shit. I'm willing to lay down any pride that I need to."

In the first and second stages, the treatment for syphilis is a simple and effective one: penicillin. Cook got ready to write a prescription, but Claire had a related body question. She once took amoxicillin and then developed some acne—*it was probably acne, but maybe a rash?*—around her mouth. She asked if that was related to the amoxicillin, and if it meant that she couldn't take penicillin because of an allergy.

"What's the worst that could happen?" she asked.

"Uh, you could die?" Cook said, sort of jokingly. It's true, but unlikely. About one in ten humans reports having an allergy to penicillin. Many fewer actually do. Doctors hand out the label prolifically just to be safe. Especially when confronted with a patient who did not photograph the rash. (If there's one piece of advice I hope to convey in this book, it is to *photograph your rashes for future reference*. Making them into a Pinterest board is optional.)

Amoxicillin is in the same family of antibiotics as penicillin, in that they work in similar ways. Doctors are trained to presume that an allergy to one member of the antibiotic family means an allergy to every member. In reality, most people labeled penicillin allergic would experience little to no untoward effects after taking amoxicillin, or even penicillin itself. Caution is born of the potentially serious complications—the few who would experience intense paroxysms of their immune systems, exploding into an antipenicillin storm that can inadvertently close the person's throat and stop the heart.

Using penicillin to treat syphilis is different from using it to treat an earache, where you take small doses by mouth for a week. Syphilis treatment involves a single injection into a person's muscle. To be effective, a whole week's dose goes into you at once. In that case, allergy symptoms would be more severe, and the penicillin regimen can't be halted.

Ultimately Claire and Cook decided that the possible-but-not-really rash was not grounds enough to forgo penicillin. Other antibiotics are effective but carry their own risks. She left with a penicillin prescription and a spring in her step. In parting, she asked me to write about how gorgeous she was. One of my colleagues in medical school was expelled after allegedly, as rumor had it, complimenting a patient's breasts during a medical exam. So, outside the realm of my medical appraisal, but because she was candid in sharing her experience purely so that others might know more than they currently do about sexually transmitted infections, she was.

How do cells from my genitals create another human's brain?

You're talking about babies?

. . . Yes, babies.

I suppose this starts as the question of how we regenerate. Red blood cells, for example, live for only ninety days, and then they die. So *why don't we run out of blood?*

He never found what he was looking for, but the histologist Alexander Maksimov imagined that we must carry around some "stem cells" from which different types of cells emerge throughout our lives. He coined the term in 1908 to refer to cells that could become other types of cells.

Even once we knew that we have stem cells in our bone marrow, this was far from explaining how a microscopic ball of cells inside a uterus could become a human with a heart and a brain and bones. It was not until 1981 that developmental biologist Gail Martin at the University of California, San Francisco, discovered that there is a type of human cell that can become any kind of body cell. These were the first known stem cells that were "pluripotent," or possessing many powers. (If ever there were a word deserving of wider use, that might be it.)

A stem cell is a blank slate. It could just as well become a pacemaker cell in your heart as it could a fingernail clipping. Brain cells could just as well have been boring gallbladder cells, storing and pumping out bile (literally rather than figuratively).

We all begin as embryos that are purely stem cells, until a hormonal milieu tells cells to develop specific structures and functions. That milieu is the key. It is the domain of the emerging field of *epigenetics,* which examines the role of environment in influencing how our genes are expressed—as in how twins with identical genes can turn out so differently, and how all humans can share 99 percent of their DNA and yet be so unique. Individuality is much less about our genes than it is about the way our genes end up working (being turned on and off, in what arrangements and magnitudes). Epigenetic effects are so powerful that cells with the exact same DNA can become nerves, bones, muscles, and so on.

But these cells do not turn back into stem cells, and they inevitably undergo senescence, the biological term for aging that connotes a slide into sickness and frailty. This is why it's so fascinating that an infant, despite being the product of two aging human bodies, has organs that show no signs of aging.

The key is that we carry cells known as germ cells in our ovaries and testicles. Those germ cells create egg and sperm cells, which combine to form stem cells. And these germ cells do not undergo senescence.

Senescence is commonly seen as the shortening of the telomeres (the caps at the ends of the chromosomes). They gradually degrade every time a cell divides. Eventually, when the telomeres are critically short, the cell can no longer divide, and it dies. The telomeres within germ cells, however, do not shorten. They contain an enzyme that rebuilds telomeres. Called telomerase, it is almost undetectable in every other cell in our bodies.

With their ever fresh chromosomes, germ cells represent a link between mortality and immortality. It seems now that if you were to announce, "I've got your key to immortality right here," and gesture to the germ cells in your groin, that might be wrong but not incorrect.

So if our bodies can and do produce cells that don't age, why don't they do that for all of our cells? (At least the ones in the face?)

And, knowing that immortal cells are physically possible, could we figure out a way to make all of our cells immortal?

And then, if we could, should we?

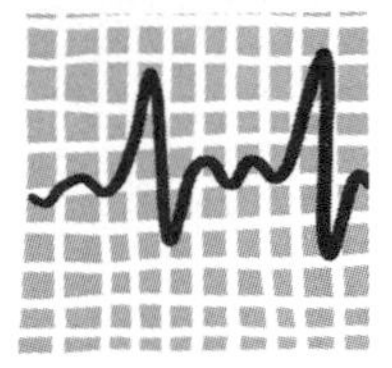
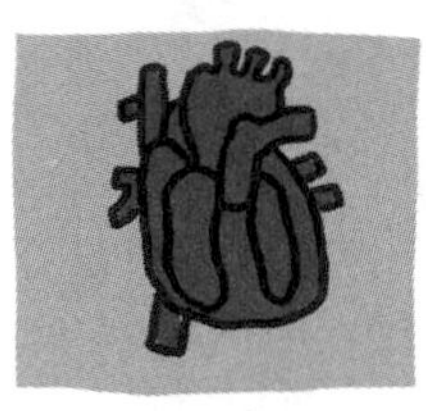

PART SIX

ENDURING

THE DYING PARTS

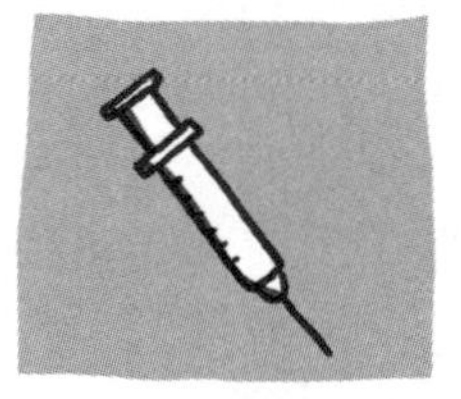

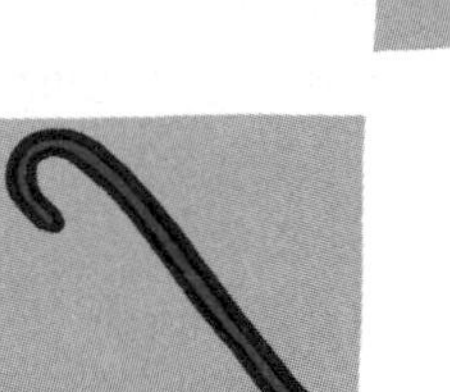

On a gray Sunday morning on Manhattan's Upper West Side, little friends swarmed Rafi Kopelan. It was still early March, so she was especially bundled, a marshmallow whose only exposed skin was a smiling pink face. She's short for seven. Eating has always been an ordeal. Swallowing food damages her esophagus, and it forms scars that make it narrow. An average person her size would have a 16-millimeter esophagus; hers gets down to 2. So every few months, she goes in for a procedure known as "balloon dilation." True to its name and similar to what's done in coronary arteries, a gastroenterologist drops a medical-grade balloon down Rafi's throat and inflates it, forcibly expanding the caliber of her esophagus. Once the balloon is deflated and removed, food should be able to pass. At least until the tissue scars and narrows again.

Rafi meandered through the festive crowd, holding her aunt's hand. Around a hundred people had come out for a charity "fun" run as winter wind whipped off the Hudson. They were there to raise money for epidermolysis bullosa research. Many knew Rafi and her family well. They wore white T-shirts for the event (called Rafi's Run), printed inside out in solidarity, since the seams that run along the inside of a T-shirt can tear Rafi's skin.

Brett Kopelan started the race five years ago. His daughter wasn't expected to live this long. Collagen is such a ubiquitous protein that when it is broken or gone, as in epidermolysis bullosa, most every part of the body suffers. When Rafi was born, he saw "no hope of a cure, a lifetime of pain, financial ruin." He set out to personally meet every researcher around the world and learn about every clinical trial in history. (This was an attainable goal.)

"These kids end up losing their hands to something called

pseudosyndactyly, which means the fingers fuse together," he described to me. "Hand release surgery opens the fingers and straightens them. But in order to do that, you have to lose some skin in the process."

Rafi also has anemia and an enlarged heart. At the University of Minnesota, Brett found John Wagner, who was doing a trial with bone marrow transplants. Seeing no better hope, the Kopelans moved from Manhattan to Minneapolis. In October 2009, Rafi became the eighth kid in the world to undergo a *fully myeloablative* bone marrow transplant. That means she received enough chemotherapy to kill off her own bone marrow's stem cells, and then, once her bone marrow was dead, Wagner infused blood from the umbilical cord of a newborn donor baby (from Germany).

"It was a godawful procedure," Brett recalls. While her immune system was gone, Rafi contracted lymphoma and almost died of pneumonia. She was in the hospital for about a year and a half. But her body did not reject the new cells. They started producing collagen VII. But, for reasons unclear, not enough of it. "We gave her a less severe form of the disease," Brett said. "We initially thought we'd have nine years with her. Maybe we've bought her a few more."

They moved back to New York, and in 2011, Brett devoted himself full time to running the Dystrophic Epidermolysis Bullosa Research Association. Some of the proceeds from Rafi's Run have gone back to the University of Minnesota, where researchers recently made an enormous breakthrough: being able to edit the genetic code in the cells of a person with epidermolysis bullosa and replace the mutated DNA segment.

Those cells have begun to produce collagen VII in normal quantities. Brett enthused over the potential for this line of research when we ate Mexican food down the street from his office near Wall Street. Known commonly as "gene therapy," it involves borrowing the technology found in retroviruses (such as HIV) to alter a person's DNA. In Rafi's case, it would mean removing the mutated gene and replacing it with one that would allow her to produce collagen VII. At the University of Minnesota, this has worked, but only in a dish in a lab. The next step is to figure out how to deliver these new cells into a person.

In early 2016, CRISPR Therapeutics—one of a few start-ups using gene therapy—partnered with the $105 billion pharmaceutical company Bayer AG to work on medication development. Their grand goal is to get gene-editing technology into people's cells wherever it is needed. If that works, it will be among the most important developments in the history of medicine. As it's supposed to go, an RNA sequence called CRISPR binds to a DNA sequence in a person, and an attached protein called Cas9 cuts that DNA sequence out. "Gene drive" technology can then be used to insert a replacement segment of DNA (probably by using viruses as delivery vehicles for replacement genes).

No one has yet been able to make CRISPR drugs that can be accurately directed to diseased cells within a living person and then, once there, correctly identify and replace the abnormal genetic code. That's a challenge that's orders of magnitude more complex. But Brett is optimistic.

"I think gene therapy is the silver bullet," he told me. For all that the Kopelans have been through, he remains sanguine on the science. "It's a very cool time in medicine."

In the meantime, though, the Kopelans deal in hope. "Going to the bathroom is a big problem. It's traumatic. We're doing prebiotics and probiotics, but who the hell knows?" Rafi still deals with daily itch, which she can't scratch without tearing her skin. Brett looked up from his burrito bowl: "If you can create a drug for itch, you're a multibillionaire."

How does my heart know to beat?

Cord Jefferson's heart woke him up at 3 a.m., vibrating his pillow by way of his mattress. The irregular racing kept up until

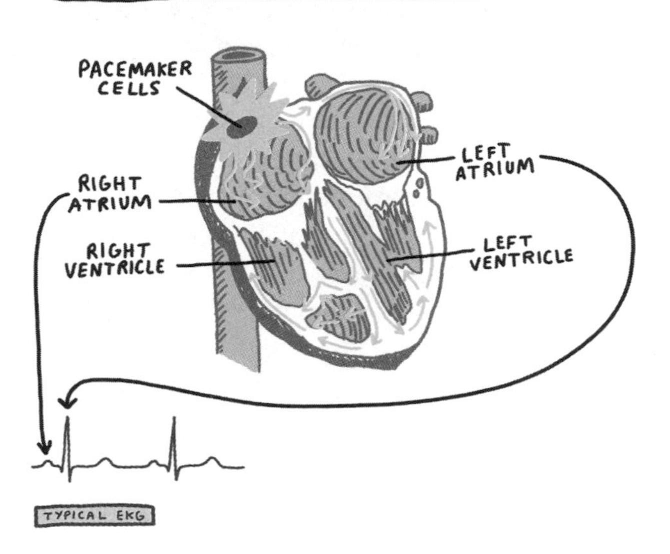

morning. He likened it to a jazz drum solo. When the sun came up, the thirty-two-year-old comedian took an Uber to an urgent care center in Brooklyn Heights. There his heart was beating 142 times per minute, more than twice what it should have been.

The doctor at the clinic recognized the irregular pattern and told Jefferson he needed to go immediately to a hospital—by ambulance, not Uber. (Uber ambulances are not yet a thing, just an idea I have that could save everyone a lot of money.)

Once at the hospital, an anesthesiologist relieved the young man of consciousness. A cardiologist shocked his chest, the medical equivalent of pressing restart on the heart.

It worked. The jolt sent Jefferson's heartbeat back into a normal pattern.

Later that day, he walked out into the sunlight with electrode burns on the skin over his beating heart, a new diagnosis of *atrial fibrillation*, and a new approach to life. He went on to write that these were the hours in which he realized his vincibility. For people born and raised in more or less good health, these first diagnoses tend to be transformative. If only for a moment, time stops feeling like a given.

Atrial fibrillation is a frequent cause of this perspective-shifting moment. In a world of spectra and gradations of normalcy, the heartbeat is one function where there is a definite normal. Anything else is an abnormal rhythm, or arrhythmia. Of the common arrhythmias, atrial fibrillation is the most serious.

It is also the most common. Defined by quivering of the atria (the upper two chambers of the heart) due to uncoordinated contraction of the individual muscle fibers (fibrils), the process now happens in around 3 percent of people—as best we know. It's impossible to say that your atrium has never once fibrillated, if only for a moment, unless you've worn a heart monitor 24/7 since the day you were born. (Have you? Call someone.) A bout of atrial fibrillation may have felt simply like a passing sense of light-headedness, anxiety, or weakness.

Even as technology to treat this extremely common malfunctioning of the heart has ostensibly advanced, the World Health Organization recently found that every year the condition causes successively more death and sickness—and unknown amounts of subtle symptoms like fatigue. Atrial fibrillation makes a person five times more likely to have a stroke, and twice as likely to die young. New cases increased by about a third in the 1990s and 2000s alone, and are projected to double in the next fifty years.

Part of that increase is due to people living longer. As we age, our older atria are more likely to fibrillate. But cardiologist Sumeet Chugh, who led the World Health Organization investigation, says the boom can't be entirely accounted for by the aging population. The heart malfunction has been linked to obesity and air pollution, among "several other factors" that come down to lifestyles and environments.

To consider why our hearts are falling out of sync like never before—and why it's killing us despite ever more impressive treatment approaches—is to consider the central paradox of modern medicine. Our hearts know to beat because of a finely tuned process that evolved over millennia, and only in the past decades have we figured out how to systematically undermine that process on such a large scale. The solutions to prevent it are before us, but instead we have created a system predicated on treating the condition—shocking and burning people's hearts to temporarily restore normalcy, at great cost and risk, most often without addressing the fundamental causes.

Our coronary arteries are so named because they run along the outside of our hearts in the shape of a crown. When we live in ways that cause these arteries to harden and fill with plaque, they narrow, allowing less blood to pass. Once a coronary artery is totally blocked, an area of the heart loses its blood supply. The heart is attacked; the muscle dies. This is the most likely way that each of us, and everyone we know, will die. More people will die this way this year than from all cancers, infectious diseases, and war combined.

All of this comes down to a focal point in our heart, a centimeter wide, made of a group of specialized cells that can generate electricity. These cells, known collectively as our pacemaker, live embedded in the wall of the right atrium, the small chamber where blood enters the heart. Every second or so, those cells create a pulse of electricity that spreads through the muscle of the heart and causes it to contract. (When people have an artificial pacemaker device implanted in their chest wall, as about one million people do every year, the device creates electrical signals that simply mimic or override the heart's pacemaker cells.)

When the muscles in the walls of our atria contract, they

dump blood into the lower two chambers. These, the ventricles, have thicker walls with three layers of muscle that can squeeze blood out into the body, all the way to the toes. (Even when you're standing on your head, the two-inch left ventricle can get blood to your toes.)

This requires not just muscle, but perfect coordination. The powerful left ventricle must squeeze at exactly the right time, in harmony with the rest of the heart. The intricate process happens about one hundred thousand times a day, thirty-five million times a year, so subtly that we don't notice until it breaks. For all that must happen for it to go right, it is amazing how rarely it goes wrong.

Even when it does malfunction, our bodies can survive most irregular patterns for a while. But if the heartbeat remains erratic—or simply too slow or too fast—people begin to die. The most serious and common reason is due to the simple fact that blood must always be moving. When it stops moving, the liquid morphs into a solid. It forms stubborn clots—which are meant to save our lives when we're injured. But at the wrong time and place, internally, they kill us. They kill us *all the time.*

Irregular rhythms leave pools of stagnant blood in the atria, which form clots that get sucked down into the ventricle, and then catapulted up into the arteries that supply the brain. The clot lodges in a vessel, blocking the flow of blood, causing brain tissue to die. (This is a stroke.)

If the clot is large enough, we may die before hitting the ground. Many prefer that to the outcome when a clot is slightly smaller or less fatally positioned, which can mean paralysis or dementia. If the clot is tiny, we might wind up with just a moment of forgetfulness or a sense of déjà vu, known as a transient ischemic attack.

Which is all to say that even when atrial fibrillation is not causing a racing heart or any symptoms, it's important that the fibrillating is quelled. The question is how.

What is sudden cardiac death?

Even if the ranking process for academic medical centers is less than scientific, it would seem to mean something that *U.S. News & World Report* has ranked the same hospital number one in the nation for cardiology and heart surgery—Cleveland Clinic—for twenty-one consecutive years.

This accolade is marketed rigorously on the Cleveland Clinic's website, which also warns visitors that the phenomenon known as sudden cardiac death is the "largest cause of natural death in the United States." It kills about 325,000 people every year. Usually these are people in their thirties and forties. But in search of an actual definition, Cleveland Clinic begs the question. "Sudden cardiac death is a sudden, unexpected death caused by loss of heart function." (You guys, why do you have a website?)

When we hear about someone just dying out of the blue, it's most often classified as sudden cardiac death. The causes can be many, but they all come down to a malfunction of the electrical system. Usually, the heart goes into spasms. The ventricles instantly begin fibrillating—which is much more serious than when the smaller atria fibrillate. At this point, the heart is moving but not "beating." You are technically dead. Your heart is not pushing blood to your brain, but there is still enough oxygen there to let you feel the heart spasm for a few seconds, long enough to mutter something and grab your chest before you lose consciousness.

As the name suggests, sudden cardiac death does tend to be fatal—unless someone right next to you has a *de*fibrillator that can shock you back to life, or knows how to operate a human heart with their hands and shoulders until it restarts or someone with a defibrillator arrives. CPR works occasionally. Between 2 percent and 16 percent of people who get CPR from a bystander—for any reason—will make it out of the hospital alive. Resuscitation is slightly more effective if you die while in the hospital, with around an 18 percent survival rate ("survival" being a catchall term for states in which the heart is beating on its own and the brain is not fully dead). And during the time that the brain has no oxygen, most people endure some degree of brain damage that can never be erased.

. . . Why did I think CPR was more effective than that?

Researchers at Duke University traced this misconception to the television show *ER*.

The 1990s surge of medical dramas began with *ER,* which bled into the CBS bastardization *Chicago Hope*—down to the casting of Adam Arkin, who looked eerily similar to *ER*'s breakout star George Clooney. Both were masters of the knowing closed-mouth grin, though Arkin never captured hearts the way that Clooney did (possibly because he lacked dimples).

ER survived Y2K and spawned other medical dramas: *House,* a Sherlock Holmes version of *ER,* and *Grey's Anatomy,* a sexy version of *ER* that I'm told is still going on today.

In the canonical 1990s hospital dramas, people died and were brought back to life with CPR at rates four times higher than in real life. The researchers at Duke watched more episodes of *ER* and *Chicago Hope* than could plausibly be considered healthy and calculated a fantastical survival rate of 75 percent. In the show, two-thirds of people left the hospital with normal brain function.

This almost never happens. Once you die, the chances of having a good life are small. Trivial as a study like that might seem—*it's just TV*—these depictions form many people's understanding of cardiac death, because we don't talk much about it otherwise. And realistic expectations of the dying process are critical to making sure we die on our own terms—as much as possible—which most Americans do not.

The TV study was important enough that it warranted publication in the *New England Journal of Medicine.* "Given the media's extraordinary influence," the authors concluded, "we could hope that the producers of television programs might recognize a civic responsibility to be more accurate. This may not happen, however."

They were correct in that last bit. Legal scholars have similarly documented a "CSI Effect," wherein fans of crime procedurals eventually find themselves on real-life juries and have unrealistic expectations of prosecutors. It seems there is a just as pervasive *ER* Effect, wherein we have unrealistic understandings of our own mortality. If *ER* reflected reality, the show would have depicted a lot of death, and even more dying. It would've been a barrage of gradual transitions from life support to nursing homes, laden with dementia and depression and laundry lists of medications passed back and forth between care facilities and hospitals, and everyone sidestepping talk of the inevitable end. The central drama would have been constant, protracted arguments with insurance companies refusing payment. It might have felt out of place in NBC's Must See TV Thursday nights after *Seinfeld* and *Friends*.

In the "real world," the best option is of course not to endure sudden cardiac death, though every year a smaller percentage of people manage that. To understand how the heart's electrical system can be repaired and revived is to understand the incentives of the most expensive health care system in the world (the United States'), which has every financial incentive to keep us alive, but little or no financial incentive to keep us well.

Why do heartbeats mess up?

Modern cardiology is built around an approach to the heart that is typified by a condition called Wolff-Parkinson-White syndrome.

A rare cause of sudden cardiac death, Wolff-Parkinson-White is named for the three cardiologists who described it in

1930. Just six years earlier, the Dutch physiologist Willem Einthoven had received the Nobel Prize for inventing the electrocardiogram (often called an "EKG" instead of "ECG," because the German is *Elektrokardiogramm*). It is still today the most common and (arguably) most valuable test available to doctors.

Immediate, cheap, and accurate, an EKG maps the electrical currents as they travel down the heart. Taped to a person's chest, electrodes detect the currents inside it and translate the signal into the physical motion of a pen (now digitized). Einthoven also established what the tracing of an electrical pattern looks like for a normal heart.

The three doctors Wolff, Parkinson, and White were early adopters of the technology, part of a new field of medicine concerned with our electrical patterns, called *electrophysiology*. The men collected their observations of abnormal heartbeats that corresponded to abnormal electrical patterns. One such abnormal EKG tracing became known as the iconic *delta wave*—now one of the first things that students learn to recognize in every medical school around the world.

The delta wave meant that the heartbeat was still originating normally—from electricity generated in the pacemaker cells of the atrium—but instead of traveling the normal route down to the ventricles, the signal was taking a shortcut. That meant it arrived at the ventricles too early, causing them to contract prematurely. This tends to happen in one in five hundred people. Most are fine and may never know it. Others die suddenly. Between those extremes, as in atrial fibrillation, people feel occasional lightheadedness or dizziness.

One of the syndrome's three eponymous discoverers, Paul Dudley White, was a Harvard professor and chief of cardiology at Massachusetts General Hospital. He was among the founders of the American Heart Association and the International Society of Cardiology. For his pioneering work against the growing global epidemic of heart disease, Lyndon Johnson awarded White the Presidential Medal of Freedom in 1964. Over the course of his illustrious career, White became convinced that people's lifestyles were at the root of most heart disease. He developed and advocated the specialty of "preventive cardiology."

For decades, the concept was essentially left behind, though, as the field raced toward treating people who already had heart disease. Cases of heart disease continued to increase. Even as treatments became more elaborate, death and suffering from the condition grew. The country with arguably the most advanced cardiologic and cardiothoracic surgical technology became the world leader in heart disease.

And in a way, despite White's advocacy of early prevention, the failure of physicians to adopt this approach can be traced to the very syndrome that now bears his name.

For fifty years, doctors were helpless in the face of a delta wave. They watched their patients deal with fainting spells and occasionally a fatal arrhythmia. In the 1960s, cardiologists began studying the phenomenon by feeding wires (called catheters) into human hearts, by way of the femoral artery in the groin. Put your finger on your upper leg where it meets your genitals, if you want to, and you should feel that artery pulsating. After inserting a needle and hollow sheath into that artery (or its accompanying vein), doctors can feed a long wire up through the vascular system and into your heart. Through that wire, cardiologists can now do many extreme procedures, including replacing heart valves and opening blocked arteries.

It was one of these wires that first burned the inside of a person's heart as well—an accident that became one of the most consequential discoveries in modern medicine.

The year was 1978, and a man had been suffering from repeated bouts of loss of consciousness due to Wolff-Parkinson-White syndrome. To study his electrical pattern more closely, doctors ran a catheter up into his heart, carrying electrodes capable of mapping current more precisely than the external electrodes used in an EKG.

It's usually a very safe procedure. But, as the Oxford car-

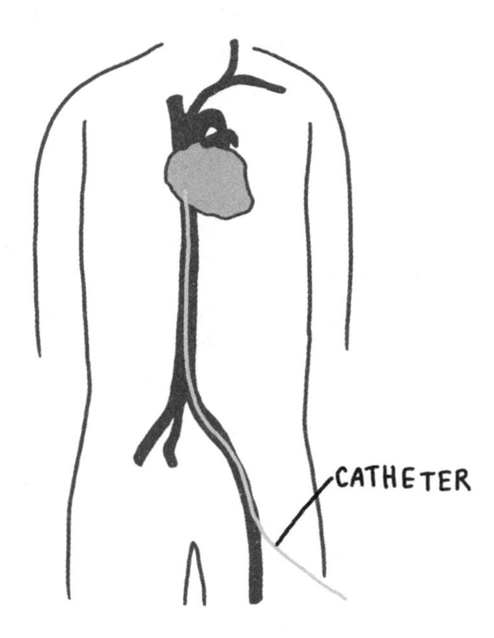

diologist Kim Rajappan recounts the event, two live electrodes came in contact with each other, "resulting in a high-voltage discharge" and "damage" to the surrounding heart tissue. The damage was severe enough that it caused the man to lose consciousness. But once he opened his eyes again, his heartbeat was totally normal. His symptoms permanently disappeared.

The doctors realized that by burning the heart, they had blocked the pathway through which the electricity normally traveled. Rajappan understatedly calls this accident "rather serendipitous." (It revolutionized cardiology.)

That fortuitous burn ignited a quest for a way to do this purposely. If doctors could strategically burn just the part of the heart that was transmitting abnormal electrical currents—like the abnormal shortcut pathway in Wolff-Parkinson-White syndrome—people could, in principle, be cured.

In San Francisco, an eager team of cardiologists procured ten dogs, ran wires into their hearts via their groins, and ran electrical currents through the wires. And the trial was successful: The doctors were able to burn all of the dogs' hearts. (Which sounds less sinister when it's called "ablation" of the tissue, as it's now known.) The dogs survived, and their heart rates were permanently slowed. The doctors deemed the technique "suitable for experiments in which heart rate control is required."

Use of the procedure spread to humans almost instantly in the early 1980s. One electrode would be stuck to a person's skin, and the other positioned inside the heart. The doctors would then use a high-voltage current to burn the tissue in between, including the chest wall. It wasn't entirely accurate, and complications were not uncommon when the doctors missed the target—which could lead to heart failure and, in some reports, what the doctors deemed "cardiac rupture."

But when it worked, it was preferable to open-heart surgery. Before ablation, the only way to destroy an abnormal pathway involved cracking open a person's sternum, stopping their heart, and opening it with a scalpel. So by the middle of the decade, electrophysiologists had launched the modern trend in "minimally invasive" procedures—where a person could be in and out of a hospital in twenty-four hours with only the tiniest of scars. To a patient, the choice between that and a cracked sternum is no choice at all. Imperfect as they were, the high-voltage DC currents could somewhat reliably cure Wolff-Parkinson-White syndrome, and spread to be used for other rhythm abnormalities.

In 1987, medical device companies began production of modern catheters, which use radio waves to heat an electrode in the tip of the catheter, allowing bestowal of a burn that is precise and accurate—and quick. The American Heart Association describes the modern practice in blunt terms to potential consumers: "Radiofrequency energy (similar to microwave heat) destroys a small area of heart tissue that is causing rapid and irregular heartbeats."

Uncomfortable as it may sound, there are no pain receptors on the inside of the heart, so people can often be conscious during the ablation. Just as staring at the sun painlessly incinerates your retinas, someone could be burning the inside of your heart right now. (Though you might notice the wire coming out of your groin.)

In people with Wolff-Parkinson-White syndrome, cardiologists can test the speed of the abnormal shortcut pathway. The faster it transmits, the higher the person's risk of sudden cardiac death, so the more reason to do an ablation. Between that syndrome and other similar arrhythmias, millions of lives have been improved, if not saved, by way of strategic burns. People are usually cured for life. *The New York Times* reported in 1998 that ablation "corrects some debilitating heart defects forever. Ninety-nine percent successful, it eliminates the need for medication and is often an alternative to open-heart surgery." Just as rates of cardiac death were soaring, ablation seemed almost too good to be true.

. . . So with all this technology, **why are more and more people dying of heart disease?**

That brings us back to the most common abnormal rhythm, atrial fibrillation (the one that affected Cord Jefferson and doubles a person's risk of premature death, quintuples their risk of stroke, and is set to continue skyrocketing).

It was in 1998, as atrial fibrillation diagnoses were in early surge, that electrophysiologists in Bordeaux, France, offered hope that another revolution was at hand. Michel Haïssaguerre and colleagues reported in the *New England Journal of Medicine* that they had successfully treated twenty-eight patients diagnosed with atrial fibrillation, by using strategic ablation of their heart tissue in areas around the pulmonary veins. These people no longer needed medication. The news spread quickly.

The practice had been so successful for other arrhythmias

that it was an easy sell for doctors to believe it could work against this increasingly common one. Almost instantly, doctors around the world began doing the procedure—burning hearts to treat atrial fibrillation.

Among them was American electrophysiologist John Mandrola. He practices now in Louisville, Kentucky, where he moved after years as director of the arrhythmia center at the VA hospital in Indianapolis. Kentucky is, like Indiana, among the most obese and least healthy states in the country. (When I rotated at the Indianapolis VA as a medical student in 2008, there was a cigarette vending machine in the lobby.)

Mandrola rode the wave and began doing ablations for atrial fibrillation in 2004. Within the decade, half of the ablation cases at his five-hundred-bed hospital in Louisville were being done for atrial fibrillation specifically. One of Mandrola's patients complained to him though when he received a bill for the ninety-minute procedure (and overnight observation in the hospital), which came to over $100,000.

Insurance companies often end up settling the bill for closer to $20,000 to $30,000. But if you didn't have insurance, Mandrola reminded me, then you would have to negotiate that yourself.

Many new operatories have already been opened around the procedure, a sound financial investment for any hospital CEO. Biosense Webster, a leading manufacturer of ablation catheters, estimates that the number of ablations done in the next five years will double. Of the nearly four hundred thousand done in 2020, according to company projections, two-thirds of them will be done for atrial fibrillation.

In a profession where it is taboo to talk of money, this degree of expenditure is common. To put a price tag on an additional day of life can lead to a slippery philosophical slope. But this case is exemplary because, even at this great a cost, it is becoming apparent that the benefit of cardiac ablation for atrial fibrillation may be very small, if it exists at all. As cardiologists have begun to look back at their decades of ablating atrial fibrillation, they have seen only *more* people dying as a result of the disease.

The influential cardiologist Sanjay Kaul, director of the Heart Rhythm Center at Cedars-Sinai in Los Angeles, agrees with Mandrola's skepticism. "The success rate has been overhyped," he told me. At this point the evidence that the procedure even reduces morbidity or mortality is "insufficient."

While some people do say they feel better after the procedure, and some have fewer episodes of atrial fibrillation, no one has yet shown that ablations actually reduce the likelihood of a stroke, or heart failure. Rather, in 2015, cardiologists at the University of Illinois actually found that the procedure itself caused acute heart failure in about one-quarter of patients, almost half of whom required hospitalization within a week afterward. As Mandrola put it, "These are not Mickey Mouse procedures."

Unlike the classic, simple ablations done for Wolff-Parkinson-White and other arrhythmias, those for atrial fibrillation often involve more than fifty burns to the heart. The key to ablation for classic electrical problems was simply finding the abnormal pathway and knocking it out. In atrial fibrillation, though, a very small percentage of people have an identifiably abnormal electrical pathway. The signals seem to come from all over. So, in a sort of shotgun approach, electrophysiologists make informed guesses about where to burn—about the source and path of the abnormal rhythm.

"For the past decade, everyone has been searching for this one target where you can ablate atrial fibrillation," Mandrola told me, only to be met with frustration about why "what we're doing doesn't seem to be working."

Since 2014, he and Kaul have slowed their role, doing fewer and fewer of the procedures. The procedure has persisted for as long as it has because not only did doctors, patients, and hospitals want it to work, but so did people with a financial interest in medical devices. Many electromagnetic catheters cost tens of thousands of dollars. Mandrola believes the rest of the field will come around eventually, though: "When policymakers take a look at the costs and the risks compared to the rare benefits, they'll have to say, 'Oh my gosh, this has to stop.'"

At the Mayo Clinic in Rochester, Minnesota, cardiologist Douglas Packer is determined to prove whether or not these procedures should be done at all. He is leading an international study of thousands of patients with atrial fibrillation. Called the CABANA trial (which stands for "catheter ablation versus antiarrhythmic drug"—sometimes these acronyms are forced), its goal is to determine if the procedure decreases rates of stroke and mortality. The trial is set to conclude in 2018. Another in Europe is ongoing.

What makes this question so emblematic of the faults in the American health care system is not just the cost and risk of the procedure, but the fact that researchers have discovered a treatment that *is* capable of preventing and treating atrial fibrillation in many cases, and is proven to extend and improve life, and yet few hospitals have moved to use it.

At Royal Adelaide Hospital in Australia, cardiologist Prash Sanders was feeling the same sense of treading water. After years as director of the University of Adelaide's Centre for Heart Rhythm Disorders, he was having underwhelming success against atrial fibrillation. He surmised, like many others, that the abnormal electrical pathway in atrial fibrillation was proving so difficult because in most patients the signals are coming from many different places at different times. This suggested that atrial fibrillation is not simply a disorder of abnormal electrical current, but a symptom of global disease of the atrium that ends up creating erratic currents. And the atrial disease itself is usually part of an even greater systemic condition of the body.

Sanders trained in France under Michel Haïssaguerre, the original champion of using ablation for atrial fibrillation. So it was unexpected that he would become a leader of the resistance. Instead of continuing to burn people in more places—maybe increasing the number of burns from fifty to one hundred?—Sanders proposed a way to turn back the clock to before the person had their atrial fibrillation. His reasoning: Almost no one younger than thirty has it. If you can reduce the things that made the atrium start firing off aberrant electrical signals to begin with—like stretching it and infiltrating it with fat—then the atrial fibrillation might get better on its own.

Sanders's team in Australia set to testing this by first procuring sheep and making them obese. The correlation between obesity and atrial fibrillation has been well proven in people and sheep alike, and as expected, many of the test sheep developed atrial fibrillation. The critical finding was that when Sanders let the sheep lose weight and move around, their atria healed and their fibrillation vanished on its own.

Encouraged, he moved to test this in people. Being as it is unethical to intentionally make people obese, it was fortunate for science that there was a long wait list to get ablations in Australia. He invited the people on the list who had atrial fibrillation (many of whom were obese) to participate in intensive programs with primary care doctors who helped them with their risk factors—not just obesity, but alcohol and tobacco use, physical activity, and sleep, among others. The results of his randomized controlled trial, published in the *Journal of the American Medical Association,* showed that atrial fibrillation and its symptoms plummeted among those patients.

Mandrola recalls hearing the results presented in a small, nearly empty room at a conference, shocked at what he thought might well be "the most important cardiology study of the decade." What most impressed him was that ultrasound images of the people's atria showed that their hearts had "remodeled" themselves, appearing visibly less fatty and stretched. When he returned to Kentucky, he wrote on his blog that this study "should change an entire way of thinking about treating people."

It made no major news. Sanders got slightly more attention when he went on to show in a 2015 study that when people do undergo ablation, adding in these lifestyle improvements increases their likelihood of being free of atrial fibrillation by six times.

This idea that atrial fibrillation is not usually a simple abnormal electrical pathway but a symptom of global disease of the heart makes sense in retrospect. Atrial fibrillation is strongly associated not just with obesity and cardiovascular disease, but with diseases as various as emphysema, diabetes, alcoholism, hyperthyroidism, and autoimmune disorders like sarcoidosis. During my medical internship in Boston, I would get paged in the middle of the night about episodes of atrial fibrillation in people with virtually any condition that landed them in the hospital. And if a person's list of diagnoses was long, it was almost a given that atrial fibrillation would be on it. Yet we regarded it as just another arrhythmia, to be managed with medications and ablations that modify the heart's electrical pathways. We needed to pull back and see the whole patient. And, really, the whole world around them.

"It's just so striking that we've been wrong about this," Mandrola said. "It's been in front of us for so long, and we haven't gotten it."

The fundamental problem may be that doctors have been treating atrial fibrillation as if it were *the* problem, when it is rather a symptom. The ablation procedure could be akin to a $100,000 Band-Aid over a gangrenous infection, when what people really need is to lose a hundred pounds, or stop smoking or drinking. But a fee-for-service system means that hospitals and doctors are paid based on how many Band-Aids they apply.

If this turns out to be a costly endeavor that—despite many

noble intentions—endangered people and distracted from more fundamental problems, it would not be the first time. Take the case of overgrown hearts. The common condition known as hypertrophic cardiomyopathy occurs when the heart muscle grows too large. This creates an unnatural buildup of pressure within the chambers. In the 1990s, doctors began treating this by implanting pacemakers. In some early reports, the pacemakers appeared to decrease that pressure. So the practice—which involves placing a metal device under the skin of a person's chest and running wires down into the muscle of the heart, overriding the heart's natural pacemaker—was widely adopted.

But by 1997, physicians at the Mayo Clinic looked back and found that in some patients "symptoms do not change, or even become worse," so long-term studies should be done before widely adopting the practice. Yet with the encouragement of the pacemaker manufacturer Medtronic, the procedure took off. Just years later, after studies concluded that only a small subset of these large-hearted people should be getting pacemakers, did the field eventually cut back.

The same thing happened recently with a procedure aimed at treating high blood pressure. In 2012, no less than Dr. Mehmet Oz cheered to his audience about what he called "a profound game changer." There was a new procedure that would help lower your blood pressure. Indeed, it had been gaining popularity for three years, since an early report in *The Lancet* said that in a small number of cases, it was as effective as a medication.

The idea of it—that people could benefit from a procedure that would ablate (with microwaves) a nerve that supplied the kidneys—was logical enough. The kidneys maintain fluid volume in the body, and increasing volume within a fixed space increases pressure. The notion of decreasing that volume by focusing on the nerves that supply the kidneys dates back to the 1940s. To treat severe high blood pressure, surgeons would open a person and sever their splanchnic nerves (which control the kidneys, among other things). The dissection often did the job of lowering the person's blood pressure—though often at the cost of impotence, occasional loss of consciousness, and "difficulty in walking."

Surgeons justified it then because the alternatives were few. High blood pressure can be quickly fatal. When it's not, it's indolently fatal. It causes strokes and heart attacks and most anything else, killing nine million people every year. In the United States alone, it affects one-third of adults and costs $46 billion. Most cases could be prevented with diet, movement, and stress reduction. But as cardiologist Michael Doumas argued in the *Lancet* in 2009, "control of hypertension remains disappointingly low," so "the need for new therapeutic strategies permits the use of interventional techniques."

Medtronic marketed its Symplicity renal denervation system for this purpose, and doctors began burning renal nerves. By 2014, doctors in more than eighty countries on four continents were doing it.

That same year, the *New England Journal of Medicine* reported the first large study of the procedure and found that it had no effect. That quarter, Medtronic lost an estimated $236 million.

The critical difference in this study was not just its size, but that it eliminated the placebo effect—which is difficult to do with a procedure. Typically, if you undergo a procedure, you know it. So the researchers had to test the renal burning by comparing it to people who underwent a "sham procedure"—the equivalent of a sugar pill. A person is taken into the operating room and anesthetized and cut, just as they would be in a normal nerve burning procedure. But nothing is done. Five hundred and thirty patients signed up to get a procedure that might be a sham. (The idea of sham procedures is itself unnerving to many. Being cut is less innocuous than taking a sugar pill. But it's the only way to fully control for the placebo effect of surgery.) Six months later, it didn't seem to matter who had the real procedure and who had the sham. Their blood pressures were the same.

The cardiologist Sanjay Kaul has said he would not be surprised if the same thing that happened with renal ablation happens with ablation for atrial fibrillation. "Patients clearly continue to have atrial fibrillation after ablation," he told me, though sometimes with fewer symptoms, "which suggests a possible placebo effect." Still, no one has compared atrial ablations to a sham procedure, so knowing whether the exorbitant procedure is better than a placebo is not yet possible. And the burden to make that distinction will ultimately fall to insurance companies, because, Kaul acknowledges, "As long as the procedure is reimbursed, the hospitals will continue to see it as a revenue opportunity."

Through this particular arrhythmia, we see the dilemma that plagues modern health care. It is a system that rewards interventions based on how elaborate they are—not on whether the use of resources makes any sense. It is a system that doesn't just ignore the concept of keeping people healthy, but has long aligned with incentives *against* it. Even if there proves to be some benefit to burning as many hearts as we are now, how is it possible to weigh that against the cost? Could the resources have saved and improved more lives if they had gone to Sanders's lifestyle solutions?

As Kaul put it, by email, "Who will sponsor such a trial! Who has a financial interest in actually keeping people healthy?"

"In medicine, there can be a sort of a feeling of futility," Mandrola told me. "'Well, okay, but this really is not going to work in America. People are not going to go for this.' They want treatment. And weight loss doesn't work immediately."

It's more than a desire for immediate fixes, though. We have also designed a medical system where experts partition themselves based on organ systems. As we live longer, the major diseases facing us are not diseases of any particular organ, but of the greater body. Atrial fibrillation is no more a disease of the heart than strokes are a disease of the brain or irritable bowel syndrome is a disease of the gut. When we fail to see the body and populations as a whole, we fail as a profession.

If atrial fibrillation is so common, do I have it?

When the World Heart Federation took notice of the rise of atrial fibrillation and the fact that it doubles a person's risk of premature death, the organization launched a campaign advocating a "DIY" pulse test—"to encourage you to detect a possible abnormality of your heart rhythm." It's straightforward. Feel your pulse, and make sure it's not erratic or faster than one hundred beats per minute. Of course, most people who get atrial fibrillation experience it only in occasional bursts. So to know for sure you'd have to be constantly feeling your pulse and counting, which is incompatible with mental health. But in spare moments, sure. Feel your pulse. Know your electricity. Feel your friends' pulses. Consider one another's mortality. Make a game of it.

Why isn't there a cure for the common cold?

The "common cold" is a set of symptoms that can be caused by many different viruses. So colds are technically many different diseases, in which our immune systems react by producing similar symptoms—runny nose, cough, lethargy, sometimes sore throats. Not only is there no cure, there's not even a test to diagnose it. That's a much bigger priority. We can now detect viruses by rapidly detecting their DNA, and someday soon this technology may come out of research labs and into clinics and hospitals. Mass spectrometry and ever better DNA sequencing can bypass culturing. We will be able to identify microbes quickly and accurately, and use precise antibiotics that kill only those harmful microbes, not more, and only when clearly necessary.

Even though cold viruses do result in a lot of missed work, they almost all run their course in a few days. We get better. Most scientists would rather work on something more urgent and dire. A quick test that could *identify* the cold viruses, though, could very well be considered urgent and dire. If doctors could know definitively that a person's symptoms were caused by a benign virus, millions of courses of antibiotics could be saved for people who actually need them. That could save millions of lives.

How do I convince my friends that their kid doesn't need antibiotics every time she gets the sniffles?

Biosis means "life," so, as a health product, the term *antibiotic* doesn't seem designed to sell. Still, over past decades the word came to mean, to many people, *cure for whatever sickness I have*.

Sniffles are almost always caused by viruses. Viruses are not technically alive, in that they are DNA wrapped in protein, so they cannot reproduce on their own. To do so, they must infect living organisms and hijack their cellular machinery. As such, most biologists do not consider them to be living organisms. Trying to kill something that is not alive with an *antilife* medication is like trying to strangle a zombie.

Worse than being ineffective against viruses, abusing antibiotics does enormous damage to all of the life that they *anti* within us. As a class of medications, antibiotics may be the greatest advancement in medical science, but every time we use one, it does damage too.

Still, the biggest danger is that with every use, we train bacteria to evolve ways to survive these antibiotics. Some people like sports, so, to use a football analogy: Teams will study footage of other teams so that the large players can knock each other over faster and harder. That way, one football team won't take the ball from the other team and achieve victory. The home team will have better success getting the ball into the end zone.

It's the same way with bacteria. When we use antibiotics, we are showing the other team our playbook. They can figure out ways around it—and not slowly.

This is already happening with, among others, the bacteria that cause gonorrhea. For years, many doctors prescribed a nonspecific, clear-cutting type of antibiotic to anyone who had a suspected case of the very common infection—even though around 75 percent of cases are completely treatable with antibiotics that specifically target gonorrhea. At the same time, while gonorrhea used to be easily treatable with fluoroquinolone antibiotics, these antibiotics were used widely and inappropriately for myriad conditions. So gonorrhea learned to survive them. Then cephalosporin antibiotics became the standard treatment for gonorrhea, but now gonorrhea is becoming resistant to those too. When I wrote about this "Super Gonorrhea" in 2012—marking the first time in its history that The *Atlantic* used an emoticon in a headline ("Here It Comes: Super Gonorrhea :-/")—some people initially thought it was sensationalistic. But I wouldn't call something Super Gonorrhea unless I believed it was worthy of serious attention. (News outlets now use the term, a vindication that is no victory.)

Unlikely as it is that your friend's child has gonorrhea, pediatricians have created similarly ominous scenarios by prescribing antibiotics without even looking into a kid's ear, and by prescribing antibiotics every time anyone gets the sniffles. Instead of protecting our playbook, we've essentially been going over to our opponents' houses and running our plays on their lawns. So antibiotic-resistant infections have become an enormous problem around the world. The former UK prime minister David Cameron recently commissioned a group of experts to quantify the global risk posed by the overuse of antibiotics. As those antibiotics lose their effectiveness, and their use leads bacteria to evolve into more virulent strains, the experts came back to Cameron with a grim forecast. By 2050, expect that antibiotic-resistant infections will kill more people every year than all cancers combined. The UK's chief medical officer, Professor Dame Sally Davies, called this an "antibiotic apocalypse" and said it should be formally registered as a civil emergency, which is the British version of an emergency.

"If we fail to act," Cameron said in response to the report, "we are looking at an almost unthinkable scenario where antibiotics no longer work and we are cast back into the dark ages of medicine." That would be because of our own hubris and shortsightedness (also civil emergencies).

This sentiment has been echoed by scientists around the globe. The Centers for Disease Control estimated that in 2016 at least twenty-three thousand Americans would die due to antibiotic-resistant microbial infections. It has also noted that while overuse in humans is an issue, by far the greatest source of abuse is when those same antibiotics are given to factory-farmed animals. Even when their animals are not sick, the proprietors of animal feedlots and slaughterhouses have long known that giving antibiotics to animals causes them to gain weight. These animal factories are reimbursed by the pound, no matter how that pound comes into being, so the practice of inducing obesity with antibiotics is good for their bottom line. The weight-gain effect seems to be a by-product of the antibiotics disrupting the animals' microbiomes, which would normally facilitate digestion and passage of food.

Meanwhile, because antibiotics are a low-margin product for pharmaceutical companies—and because their existing products are already selling *very well*—the industry has not invested in research and development of new products. The supply of new antibiotics that can kill the resistant microbes has been minimal. So antibiotics are wasted on people and animals who do not need them, while those who do need them have inadequate access.

So my advice is to tell your friends that antibiotics will deplete their kids' gut microbes. The nice thing about this approach is that it is the truth. It's an immediate reminder that antibiotics are not benign. If the swell in marketing and sales of probiotic products to parents is any indication, people are beginning to appreciate that they want their kids to have robust microbiomes. And the personal microbiome approach tends to work better than any grander, society-level argument.

Three hundred million future deaths due to overuse of antibiotics? *I know I should care.* But do not mess with my kid's microbiome.

Penicillin is made of mold?

A century ago, the Scotsman Alexander Fleming harvested the secretions of the mold *Penicillium,* which he initially called "mould juice," and then later "penicillin." At first he did not realize that his mold juice held such power. Before 1928, he was just another guy who collected mold juice. As he would later recall, "When I woke up just after dawn on September 28, 1928, I certainly didn't plan to revolutionize all medicine by discovering the world's first antibiotic, or bacterial killer. [*JH: This guy gets it.*] But that's exactly what I did." Some pharmaceutical companies picked up Fleming's mic, dusted it off, and figured out a way to create a synthetic version of penicillin, which today contains no mold.

If my mucus is green, it means I need antibiotics?

No, the color of mucus can't tell us whether an infection is bacterial or viral. It can tell us only what color our mucus is. That can be fodder for conversation, as a last resort.

What causes cancer?

Marston Linehan thought he discovered "the kidney cancer gene" in 1982. He was thoroughly incorrect, but productively so.

When the towering, soft-spoken doctor went through his surgical training the decade prior, "kidney cancer" was treated as one disease. For every person with a kidney tumor, Linehan recalls, "We did the same operation, we gave them the same drugs."

The drugs and operations worked almost never. If you had a kidney tumor that was more than three centimeters wide (about an inch), the chance you'd be alive in two years was around 20 percent. In Linehan's clinical language, he recalled to me "outcomes were poor."

Though the concept of genes causing cancers was in its infancy, Linehan had a feeling that there was a target in the code. This was less than three decades after Rosalind Franklin and Raymond Gosling had published the first X-ray images of DNA, which established the double helix structure, the winding instructions for building cells. Just four different chemicals (arranged as "base pairs") determined all of the difference in human life. It was all arranged into a mere twenty thousand or so genes. Linehan was among the many researchers who believed that one of those genes—a discrete entity in this minute and clearly quantifiable helix—held salvation. If he could find the gene, he could understand "the cancer pathway."

This was also many years before the human genome would be sequenced, offering a road map on which to base genetic differences. "People said to me in those days, 'What are you doing?'" he recalls. But sure enough, once he started looking at patients who had kidney cancer, he found an abnormality on a chromosome. Linehan, now chief of urologic surgery at the National Cancer Institute in Bethesda, Maryland, attributes it to the brashness of youth that he believed he had found *the* kidney cancer gene.

"We now know kidney cancer is caused by *at least* sixteen different genes," he explained when we met on a winter day in Washington. Several of those genes he personally discovered. Different forms of cancers are caused by permutations of genes and their patterns of expression, in concert with a person's life-style and environment, in orchestrations beyond his imagination thirty years ago. But Linehan's work on the few cancers that *are* predictably caused by single genes in predictable ways has helped illuminate the mystery of what cancer *is.*

On April 23, 1987, Linehan removed a young girl's kidney, and she lived until New Year's Day of the following year, he recalls with notable specificity. He took the cancer cells and kept them alive in his lab, where his colleagues compared the DNA to that of 4,312 patients with similar tumors. In 1996, working together with a group in England, the team found the gene, now called *VHL,* that causes a common type of kidney cancer known as clear cell.

In his lab, teams now grow *VHL* and other cancer cell lines in dishes, and they keep seven hundred cages of mice that are genetically engineered to get different kinds of kidney cancers. Still, he says nothing compares to studying how these tumors actually work in people in the real world, and what he has found by studying patterns in families.

For instance, in 1989, a young woman from Charlottesville, Virginia, came up to Bethesda to see Linehan. He removed an enormous kidney tumor and she died nonetheless, seven months later. The following year, her mother died of what appeared to be the same type of cancer. Linehan scrutinized slices of the young woman's tumor under a microscope, but couldn't tell what kind of cancer it was. His pathologist colleagues said they had never seen it before. The team kept analyzing, and in 2001 they found a mutation linked to a rare syndrome that he called hereditary leiomyomatosis and renal cell carcinoma (HLRCC), in which people get aggressive kidney tumors called papillary carcinomas, among other types of tumors. About one hundred families worldwide are yet known to have the mutation.

"There are things in life you wish you could do over," he said, his sharp gray-blue eyes squinting off in the distance over my shoulder. "We didn't locate her family. I wish I had driven to Charlottesville and found the chief of police and said, 'You've got to help me find this family.'" Because when he did locate them again, eighteen years later, her brother, sister, and aunt had died.

After looking then at the pathways within cells that lead to different kinds of tumors, it became clear to Linehan that not only were they coming from different genes, but that they were very different diseases. Each type of tumor was the result of a different problem at a different point in the cell's metabolic pathway. To treat them all the same way was nonsense. (With this understanding, to refer to "curing cancer" is about as vague as "curing infection.")

Thinking of cancers as metabolic diseases isn't a new idea, but one that had fallen out of style for decades. In 1931, it was the Nobel Prize–winning idea of physiologist Otto Warburg. Normally, mitochondria within cells produce energy by oxidation of the chemical pyruvate into adenosine triphosphate (ATP). In the absence of oxygen, though, our cells also have the ability to produce energy by fermenting glucose, as we do during strenuous exercise. It's a backup mechanism for a body in peril. Warburg showed that cancers grew when cells switched to this method full time. Like cells that were constantly in fight-or-flight mode, they tended to outgrow other cells and bulge into tumors. To procure fuel (glucose), they fed off adjacent tissues.

"His work was kind of suppressed, for whatever reason," said Linehan, "and people didn't really understand how powerful and important it was—until about twenty years ago, really."

Understanding cancer as a metabolic disease of cells means that we can design medications that target different points in metabolic pathways—different enzymes and coenzymes that are overactive or underactive.

The gene that Linehan linked to HLRCC, for example, codes for an enzyme that is part of the metabolic pathway known as the Krebs cycle (named for Hans Krebs, who worked in Warburg's lab). Just as Warburg hypothesized, the mutation allows these aggressive cancers to shift away from the usual way of producing energy (oxidative phosphorylation in mitochondria) and into the high-gear process (aerobic glycolysis). It's as if these cells are constantly in fight-or-flight mode. This gives the tumor cells an advantage in terms of their ability to grow quickly. But this difference also serves as a target—a pathway that can be disrupted using medications, thereby killing only the abnormal (cancer) cells.

Linehan has had success in this already. An understanding of genetics led to this success, even though he will never find his "kidney cancer gene." The *VHL* gene is responsible for many functions besides those that lead to kidney tumors, and most kidney tumors are not caused by the *VHL* gene. "Kidney cancer" turned out to refer to innumerable different diseases grouped together based on crude understandings of cancer as simply *growths in certain places.*

Cancers arise by way of innumerable factors in the environment, from sunlight to smoking, as they interact with both the DNA bundled in our chromosomes, and with the pathways that translate that DNA into proteins (and thus into life). "It could also be that there are genes that cause cancers to originate, and other genes involved in the cancer spreading," notes Linehan. Sometimes the DNA is set to form a cancer regardless of the environment, and sometimes the cancer is caused by the environment—and every combination in between.

People laughed off thirty-one-year-old Peyton Rous in 1910 when he claimed to have found a virus that could give his chickens cancer. It seemed at first that an "infectious" theory of cancer was at odds with the "inherited" theory that was then accepted as truth, or for that matter with developing theories about the environmental causes of cancers. But it was true enough that within just two weeks of his exposing them to a virus, Rous's chickens had cancer. He couldn't explain how it worked either. He called his discovery Rous sarcoma virus (a study in psychology, naming a cancer-causing virus after oneself). Critics of his theory—that cancer can be caused by an infection—were served fifty-six years later, when he received a Nobel Prize.

It was not until 1979 that the laboratory of virologist Robert Gallo (one of the scientists who discovered HIV) found the first virus known to cause cancer in humans, HTLV-1—a virus that not only can cause infection but that works its way into human DNA. Then observations of viruses causing cancers came pouring in: The Epstein-Barr virus (which causes mono) can also cause B-cell lymphoma and nasopharyngeal cancer. Hepatitis C virus causes liver cancer. Herpes 8 virus causes Kaposi's sarcoma. And the most widely relevant of all was the discovery that around 80 percent of cervical cancers are caused by human papilloma virus (HPV). The HPV vaccine prevents these cancers and is easily accessible, yet still many people don't get it (mainly because it involves sort of talking about sex).

It has now become clear that infection, environment, and genetic explanations for cancer are not at odds, but all part of one elaborate, unified theory. These factors collectively alter—and maintain—the mechanisms by which our cells use energy and divide. The only common factor among the diseases we call *cancers* is that they involve cells that have abnormal metabolic pathways, allowing them to divide and grow more quickly or efficiently than noncancer cells.

In 2015, Linehan and four hundred colleagues around the world completed a project called the Cancer Genome Atlas, a collection of the genomes of many cancers. The purpose of the atlas is to allow people to pinpoint the genetic differences between cancerous cells and noncancerous cells in the thousands of known variations in metabolic pathways that are known to make cells malignant, and to tailor treatments that interrupt the metabolic pathways of the cancer cells without harming the noncancer cells. It is the beginning not of an enormous puzzle, but of many enormous puzzles—to fit these genomes into pictures that involve viruses and lifestyles. It is trending rapidly toward a landscape where every individual person's pathology is considered unique.

If I lost my nose, could science rebuild one for me?

People with liver and kidney failure—even heart and lung failure—can now live for decades with the help of an organ that they didn't have when they were born. But despite millions of organ donors around the world, there is a constant organ shortage. Paying people to donate is an idea often entertained, but that ultimately leads to a system wherein wealthy people literally harvest the bodies of those less wealthy.

So we rely on chance. Major medical centers have transplant teams on call around the clock, with helicopters ready to fly to procure organs whenever someone presents them to the market. Transplanted organs from donors are regularly rejected by the bodies of recipients. This is a costly and unreliable system that leaves terminally ill people relying on the hope of availability of an organ that could save their life. But this may soon not be the only option.

It was because Gail Martin isolated the first stem cells from rat embryos in 1981 that she called them *embryonic* stem cells. Still today, the term *stem cells* conjures *embryos* for many people, and with that, important ethical dilemmas. But because it's not 1981, we know now that fetuses aren't the only places to find stem cells. They come from the bone marrow of adults, and from the amniotic fluid inside a pregnant uterus (which can be sampled innocuously). Most non-1981 of all, we can now make stem cells out of regular old cells. The concept of *induced* pluripotent stem cells is among the most important in medicine.

The joy of being a ball of stem cells with unlimited potential is fleeting, and for us imperceptible, in that it predates the existence of a brain. Soon the world exerts its forces on us, as cells, and insists that we form a tube that will become a spinal cord, and a sac that will become a head that will contain a brain, and little rows of cartilage that will ossify and fill with calcium to become hard bones supporting muscles fed by tubes filled with red cells that use iron to carry oxygen. While the cells that would come to become our bodies might all one day have been capa-

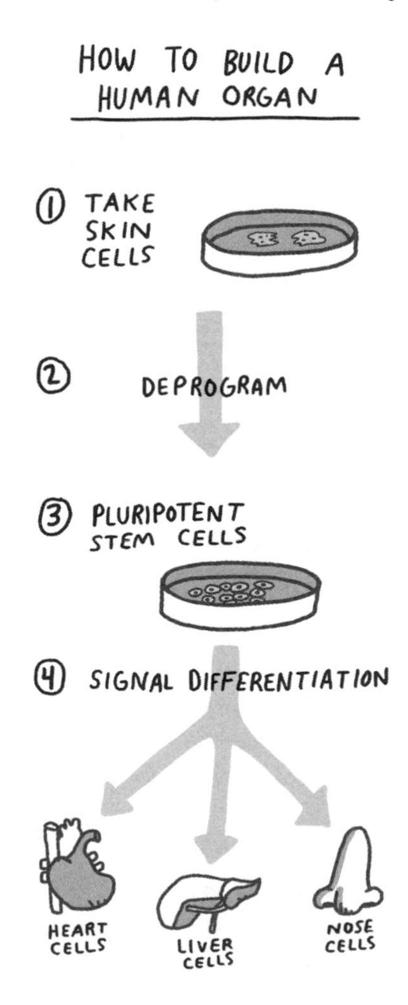

ble of becoming anything, once they become something, it is extremely difficult to change them back. But it is not impossible.

Using skin, for example, scientists have been able to turn cells back into stem cells. That stem cell can then become something else, like a liver cell. The idea is, essentially, that any part of us could become anything. All of our cells have the same chromosomes; muscle and kidney cells are different only because of epigenetic information and imprinting. By resetting that process, differentiated cells can once again become pluripotent.

But can that liver cell, made from a skin cell, then be used to build an entire liver, to replace a person's own failing liver? Could we build an entire nose? Maybe. In late 2015, researchers reported in *Nature* that by way of *induced pluripotent stem cells,* the team had turned skin cells into mini kidneys. They were not fully formed kidneys, but partially formed *organoids*. And when a stem cell that contains a person's own DNA is used to build an organ, the person's body should readily integrate and accept that organ, without the danger of rejection that comes with transplantation. A major step toward saving a lot of lives.

Is aging inevitable?

Aging can't be defined by gray hair or wrinkles or hardened arteries, because these things are common, but not universal. Steve Martin's hair went gray in his mid-thirties. I was thirty-two and *almost* refused admission to an R-rated movie when I didn't have my license with me.

What we have in common is at a microscopic level: the degradation of the chromosomes within our cells. This happens to all of us. The theater could have taken a sample of any of my skin cells and seen this with any confocal scanning microscope.

So a key distinction in terms is that aging (bodily changes over time) is different from senescence (progressive impairment of the body over time). We often accept that aging involves decline and, likely, disease. As time passes, if we live long enough, our bodies will inevitably grind to a halt.

At least this is the traditional understanding. A British technologist named Aubrey de Grey, for one, travels the world urging people to reexamine it. He claims that almost no one knows what aging really means. (Except, of course, himself.)

De Grey has the look of Methuselah in middle age, a beard down to his male nipples and a graying ponytail as far down his back. After a 2005 lecture in Oxford, England, an audience member asked him why, if he is so against aging, he makes himself look like an old man? "Because I am an old man," he said flatly, slouching in a white T-shirt and jeans of another era. "I am actually 158 years old."

He appears to be joking, though he has said the same thing in at least one other public forum. But de Grey is serious when he says that many people alive today are going to live to age 1,000 or more. Once it is possible to "eliminate" the relationship between aging and disease, he believes, life spans will grow rapidly. By his calculations, the first person who will live to 1,000 will have been born about ten years after the first 150-year-old.

The idea is not entirely unprecedented. American lobsters and hydra are among the animals that seem to be "biologically immortal," meaning that while they can be killed, as when boiled alive by humans, lobsters do not die of age-related causes. They do not *age,* in the sense that their physical capacities do not diminish with time. Why should humans not achieve the same capability?

For this phenomenon, neurobiologist Caleb Finch coined the term *negligible senescence*. He suggests that we could reach a place where aging is no longer a factor in the quality of our lives. De Grey then took the idea public, becoming a self-described evangelist for the notion that senescence need not be accepted as inevitable—his goal being "to make people realize that they are in a trance." In 2009, despite no "formal education" in science, he founded and became chief science officer of a charity "dedicated to combating the aging process." Called the Strategies for Engineered Negligible Senescence Research Foundation, the California lab has received funding from PayPal founder and tech visionary Peter Thiel (also of Sprayable Sleep). In Silicon Valley, where hubris is richly rewarded, de Grey runs a foundation predicated on extending human life indefinitely.

As most people think of aging today, it is not some bodily process that can be "cured." Yet in wealthy countries more people every year die of what are widely known as "age-related" diseases: Alzheimer's, Parkinson's, most cancers, and cardiovascular disease. (The SENS Research Foundation assures readers of its FAQ section that although age-related diseases are inevitable, at least for now, "That doesn't mean taking care of yourself is worthless.")

Here traditional research into treating and preventing diseases is merging with the avant-garde domain of *antiaging*.

The idea begins with the fact that millions of our cells divide every day. During that process, they accumulate damage in their DNA. A proper cell knows that at some point it should self-destruct. Others turn into tumors instead. Still others go quietly into the night and neither die nor continue dividing. These are the cells that are called *senescent,* and they are the target of most antiaging research.

De Grey is far from the first gerontologist (the name for the study of aging), but he has been the most insistent that age-related diseases will not be cured unless the aging process is *itself* interrupted. Where he stands out from other public thinkers is in insisting that we consider aging to be a disease process. One that is initially harmless, but eventually overwhelms us. Senescence is a "normal" part of life only from a cultural perspective, he argues. At a cellular level, it is the result of missteps. To attempt to prevent or undo those missteps is not some narcissistic fantasy, but very much in keeping with the fundamental concept of basic biomedical science.

Yet because the FDA does not consider aging to be a medical

condition or pathological process, the products that are sold to "combat" or "prevent" aging are today not regulated as pharmaceuticals. It's difficult to know what to make of the products on the market because they exist in the wild world of supplements.

There, many roads lead back to the work of Harvard researcher David Sinclair, who in 2004 came to prominence with his discovery of enzymes called sirtuins, which are involved in energy production in the mitochondria of our cells. Increasing their activity has been shown to extend life in animals (worms and mice) by a mechanism that is not yet clear. One way to increase sirtuin activity is by restricting food intake. But this is boring advice, and straight-up off the table for a lot of people. So supplement companies are attempting to create a pill, one that could stimulate the sirtuin enzymes. (Sinclair started a company called Sirtis, which he sold to the multinational pharmaceutical company GlaxoSmithKline for $720 million.)

In 2013, Sinclair reported the discovery of another antiaging compound that has also gotten the attention of not just commercial interests, but decorated scientists. Nicotinamide adenine dinucleotide (NAD) is a coenzyme, like most of the chemicals we call vitamins. Sinclair found that mice who consumed a chemical that could be metabolized into NAD ended up with younger-looking tissues. Colleagues quickly got into the business of selling this NAD-producing compound to humans, launching a supplement company in 2015 called Elysium. The venture-capital-backed start-up stands out to me among supplement companies in that it has the support of Sinclair's eminent mentor at MIT, Lenny Guarente, as well as six Nobel laureates and other respected researchers whom I know to be meticulous and judicious (like the dean of the Tufts University Friedman School of Nutrition Science and Policy, Dariush Mozaffarian).

Yet all that we know of NAD in people is that it is important to the energy-producing reactions inside our cells, and that our bodies' NAD levels decline with age. That might mean the substance plays a role in aging. It might mean that eating this substance will stall or reverse the aging process, with no untoward effects. But these *mights* are big, gaping ones. Elysium represents a dramatic turn from rodent studies to commercialization of a drug for humans. Their motto is "Optimize Your Health," and yet the fine legal print below their $60 product reads, as on all supplements, "This product is not intended to diagnose, treat, cure or prevent any disease."

In another curious aging discovery in 2016, scientists at the Mayo Clinic were able to keep mice looking young by clearing out their senescent cells altogether. Darren Baker and Jan van Deursen found that senescent cells carry a protein known as p16—so, logically, the scientists created a drug that would kill all cells that contained p16. Mice that received the treatment just twice per week ended up looking clearly younger and thinner than their birth mates, their hearts and kidneys were healthier, and they developed fewer cataracts. The animals still got sick eventually, but their morbidity was compressed into the end of their lives—as opposed to being strung out over prolonged, frail final chapters as it is for humans.

"If this paper is right, suddenly you have a way of taking an old organism and making it physiologically younger," University of North Carolina professor of medicine and genetics Norman Sharpless told *The Atlantic*. "If it's correct, without wanting to be too hyperbolic, it's one of the more important aging discoveries ever."

And across the globe, other researchers are working on other targets in the aging process. De Grey's SENS Research Foundation has focused on the accumulation of "age-related garbage" that our cells can't break down. In heart disease, for example, white blood cells attempt to break down oxidized cholesterol. They can't, so they end up full of oxidized cholesterol, at which point we call them foam cells, which are part of atherosclerotic hardening of the arteries. How, de Grey wondered, could we help our cells break down oxidized cholesterol? After considering how efficiently bacteria break down human bodies after death, he reasoned that a bacterium may be able to help us clear out things that our bodies normally don't. Eventually his team was able to find a bacterium that can break down one type of oxidized cholesterol. In a 2013 TEDx talk, he promised that this would lead to "a far, far more powerful therapy for cardiovascular disease than anything that exists today."

That's optimistic, and de Grey's claims about living to a thousand years old in the near future are well beyond those of any mortal person I've spoken to. But his quest calls to attention the ethical implications of the work being done at attainable levels—not just by reasonable people working to understand how cells age, but by people working to understand age-related cancers and dementias. Considering a scenario where we *do* live to a thousand, the immediately evident dilemma is not whether it is possible to render senescence negligible, but whether it is cool. And, without getting too syrupy here, it forces us to reexamine priorities with regard to health and longevity. What do we really want out of this?

Humans have existed for one hundred thousand years, and our life expectancies have already doubled in the last 0.01 percent of that time. In the United States, life expectancy is now 78.7 years; in 1900 it was 46.3. Accordingly, over the last two centuries the world's population has grown from one billion to seven billion. This growth can't plausibly continue. Earth will run out of food and energy as it becomes ever less habitable by way of the effects of sustaining that population.

When I was studying and practicing medicine—and contemplating stopping—I thought about this a lot. At what point is the work of treating disease and extending lives actually unwittingly precipitating the end of humankind?

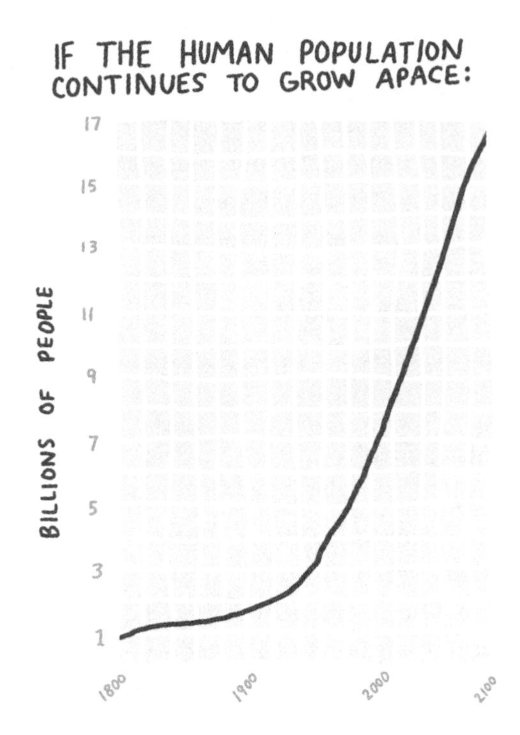

And yet if you go around a hospital asking your colleagues this, *you're* the weird one?

Why does skin become translucent with age?

As we get older, our literal skin gets thinner. Keratin breaks down and cross-links with elastin. If it gets thin enough, our veins underneath leave us with a bluish hue. The blood is still red, just a deeper shade than oxygenated blood in our arteries. Veins do look bluish when seen through your skin and subcutaneous tissue (which is how you normally see them, hopefully). But that's only because the blue wavelengths are the ones that manage to make it to your retinas. The same blood that supplies our eyes and brains deceives them. The thinnest skin is around our eyes, which is why the bags under our eyes can look dark. Theoretically if the skin were thin enough, we could have big red bloody-looking bags under our eyes. That would be much worse. Depending on what you're going for.

Is life long enough?

On a frigid Manhattan evening in February 2016, four of the world's outspoken voices on the future of aging gathered to publically debate when people ought to die. Their challenge was to convince a voting audience to decide yes or no to a question that would prove fundamental to the continued existence of the species: "Are life spans long enough?"

All were white men in suits, fittingly representative of the people afforded the opportunity to discuss the dilemmas of longevity. This time even Aubrey de Grey wore a suit, professing to the auditorium that he believes that "defeating aging is the most important challenge facing humanity." He gestured to the packed audience: "I'm glad to see that the people of New York seem to agree."

This was a debate, though, so many people did not agree. They were there because they were concerned by the problems inherent with certain people attaining the ability to radically extend their lives, while others continue to sicken and die apace. De Grey's debating opponent was Paul Root Wolpe, the first senior bioethicist for NASA, as well as a founder of the discipline called neuroethics, which examines "the social, legal, ethical and policy implications of advances in neuroscience."

The question of whether we can live to two hundred (or one thousand) is, first, the question of whether we should. Wolpe warned against embracing life-prolonging technology without considering how that technology will change what it means to be human. His debate partner Ian Ground, a philosopher at the University of Newcastle whose tie was already loosened during the opening remarks, elaborated, "There are some goods which we might say we want more of, but they are in fact intrinsically finite. I might say, 'I don't want this movie I'm enjoying to end,' and I'm really sad when the credits roll. But that doesn't mean I want to see movies that never have endings—and therefore no middles or beginnings. For then they wouldn't be movies."

Just as the central question of the life span debate was posed aloud by the moderator, Wolpe interjected with the important distinction that while *life expectancy* has increased, *life span*—the age of the longest-lived people—has really not. That is, people have lived into their nineties—and even into their hundreds—for centuries. It's simply more common today. Gains made in life expectancy have come from preventing and treating diseases that cause premature death, and this bodes well for more people every year living to 100 or 110. But it is *not* a logical extension that people will come to live to 200. As Wolpe put it, "It seems to be programmed into us that we can't live any longer than that."

Undoing that programming is the domain of Brian Kennedy. At the Buck Institute for Research on Aging in Novato, California, he studies cellular pathways that influence longevity. On that February evening, he was paired with de Grey. When asked if people should live longer, though, Kennedy's answer was equivocal.

"If I'm eighty years old, and I'm having trouble getting out of bed, and I'm taking twenty pills a day and I'm in pain all the time and I can't get out of the house," he said, as if considering it for the first time, "then maybe I don't want to live longer."

As president of an institute for research on aging, it could not be the first time Kennedy had considered this. It seemed, rather, a deliberate exercise in taking people through the thought process. Aging experts know well the visceral reactions to suggesting that any life is ever worth allowing to come to an end. When ethicists tried to address end-of-life health care during the drafting of the Affordable Care Act, the very idea was easily mangled and politicized as government overreach ("death panels"). No one ever proposed or wanted "death panels," but the ignorance was easy enough to sow. Meanwhile, the opposite extreme—where every medical technology is blindly applied to keep every person's heart beating in every situation through any cost to society or suffering to this person—is no more ethical. Yet it is our default. Kennedy seemed to realize that even questioning this default must be approached with delicate precision.

One rule is to avoid talking about money, as this seems like putting a price tag on human life. The best that can be cited are broad statistics, like that the United States already spends 19 percent of its GDP on health care, $3 trillion, and most of that is spent in the last six months of people's lives. These expenses are all poised to increase—an enormous investment in years of sickness.

"We're trying to wait till you get sick, and then spend a fortune treating you and trying to make you better," as Kennedy put it. "And if you look at the chronic diseases of aging, we're being very ineffective at that."

He argued that the goal, then, should not be increasing life span but *health span,* which he defined as "the period of time when you're disease-free, mostly at least, and when you're still highly functional." On this de Grey, Wolpe, and Kennedy agree. While life expectancy is increasing about one year in every four, health span is not going up at anywhere near the same rate. We are living longer, but an ever greater percentage of those years is spent at least modestly sick. Were we to focus on prevention even half as much as we focus on treatment, we would grow health span considerably.

It is through this lens that de Grey sees aging not as a harmless process, but as the primary risk factor in many of our biggest diseases: cardiovascular disease, diabetes, most forms of cancer, "all the neurodegenerative syndromes you're scared of, like Alzheimer's, macular degeneration, cataracts—I can go on."

He contended that because 150,000 people die every day, and many of the deaths are related to aging, it would be *morally irresponsible* to continue regarding aging as a normal process.

Ground countered that believing we have a right to an indefinitely long life amounts to believing we have the right to be other than human. "You can think that, you know, being human sucks, especially the dying bit," he said, to laughter. This is then no longer humanism, but posthumanism or transhumanism. "You

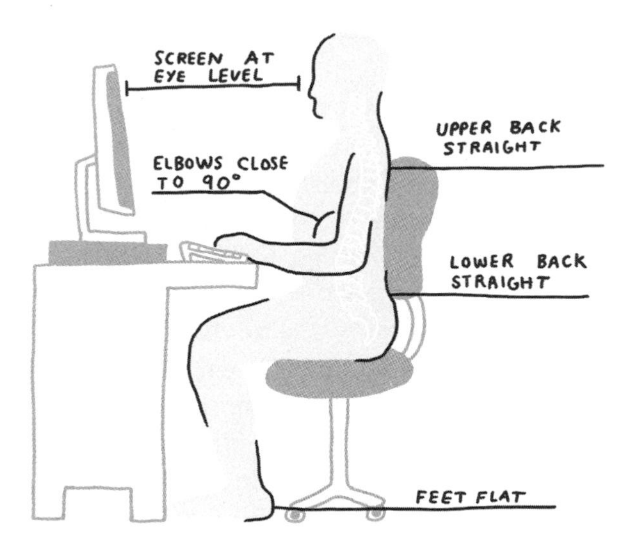

can want there to be elves, or cyborgs, or computer programs, but none of those can be *you*."

Of course, we are constantly altering what it means to be human. It just happens so slowly that we take it for granted. With artificial knees, contact lenses, and smartphones—as with the invention of shoes and chairs—cultures change as our bodies merge with technology. And as this progress brings about a longer-lived society, that society fundamentally changes.

For one, it stands to become less progressive. Older and older people accrue greater and greater wealth, further exacerbating inequality. If the World War I generation and the Civil War generation were still alive today, Wolpe posited, "Do you really think that we would have civil rights in this country? Gay marriage?"

The political ramifications of aging, and the gaps created in the process, are real concerns even to those in power today. As President Obama told me in 2016, "A democracy can't function if it's not just that some people have a bigger house or a bigger car, but they are living twenty or thirty years longer than people who don't have money. That's not a healthy society, and that's not sustainable."

The fact that people tend to become more politically conservative as they age could come down to synaptic pruning and a person's subsequent neuronal disposition. Through the changes that come with normal aging of the brain, we become particular people who think in particular ways, living lives around narratives about ourselves. As the world around us changes and our neural pathways become less prone to changing, as Ground put it, life becomes "a sense-making business."

For some of us, life already is. Even the countries with the shortest life spans today have longer life spans than the countries with the longest life spans did in 1800. In Japan, already 40 percent of the population is over the age of sixty-five. This has fundamentally changed Japanese society. There is a crisis of worker shortages, and health care costs are exploding.

Even if, somehow, our life expectancies stopped increasing, the next few generations are set to confront serious difficulty producing enough food to feed themselves. Ever more intensive agriculture (especially animal agriculture) will only warm the planet more quickly. Bill Gates has devoted much of his work to encouraging people to engage with the decision that will soon be before us: Continue to live longer lives, or reproduce less? Short of colonizing another planet, we cannot do both. (I didn't dream in medical school that I would be writing about colonizing another planet as a matter of population health.)

Interplanetary travel didn't come up that night in New York (though I'd be shocked if de Grey hasn't given it serious thought). He does concede that we will have to decide between a high birth rate and long lives. His argument for living to ten thousand, then, is hypothetical: the decision to forgo developing life-prolonging therapies so that future generations can continue having kids is not ours to make.

Of course, right now, people want kids *and* long lives. And in the end, the audience in New York voted that life spans were *not* long enough.

If even the wealthy, educated Manhattanites who recreationally listen to white men debate immortality can't be convinced that life spans are long enough, then who can? One way to get over this, maybe, is to think back to our cells. People are mortal only when we think of ourselves as individuals. It's not just symbolism—our germ cells actually become the cells of innumerable subsequent generations of people. As a species, we have the capacity to continue to produce more human cells indefinitely (in the form of babies). While the cells that constitute my body will someday not be alive, others from it will (assuming that I find a sexual mate). The body of humans as a whole is, like lobsters and possibly Aubrey de Grey, already biologically immortal. All we have to do is not screw that up.

Can I really die from popping a pimple on my nose?

It would be an extremely rare occurrence. The veins that drain the blood from your face can spread infection (every zit is a tiny, self-contained island of bacterial infection) into a part of the skull that also drains blood from the brain and cause a clot. That's called *cavernous sinus thrombosis*. It used to be invariably fatal, but since the advent of antibiotics, about two-thirds of people survive. And it's not usually caused by acne, but from some other more serious infection or blood-clotting disorder. But still, some dermatologists refer to the nose, the upper lip, and the skin in between as the "danger triangle." The more practical infection to be concerned about when meddling in that area is the flu, which is usually transmitted not by directly ingesting someone else's sneeze, but by rubbing one's eye, picking one's nose, or otherwise touching one's face. The number of people who have had serious infections from popping zits pales compared to the number of people who die every year from the flu, which is about half a million. Most people touch their face about four times per hour. If we never touched our faces, we'd never have to wash our hands. Theoretically. But you can't stop touching your face. Try it.

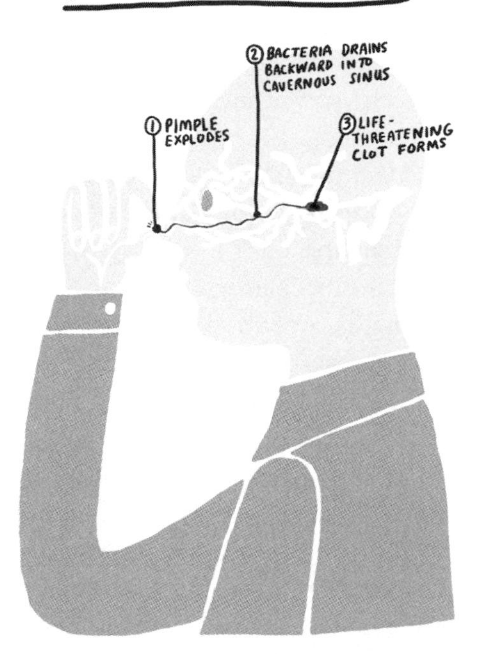

What is rigor mortis?

The default state of human muscles is rigid. They are only malleable in life because they are capable of converting chemical energy into kinetic energy. Unlike joints, which would allow for movement in whatever direction gravity determined even after death, every movement of a muscle fiber in a living person requires energy. The dead body is no longer in the business of converting food to adenosine triphosphate, so the muscles are not easily manipulated. Since it requires energy just to keep muscles malleable, we are all burning energy even when sitting as still as possible.

What happens to my body when I die?

When Benjamin Franklin was asked, in 1757, what he thought of the threat posed by Halley's Comet—which astronomers believed might very well hit Earth and end all life—he said the eighteenth-century equivalent of "meh." Earth is just one of "an infinite number of worlds under the divine government." This was a century before the idea of a multiverse—the model today accepted among many physicists which proposes that all physically possible realities are occurring across infinite universes.

If that's true and free will is an illusion, then I could end every answer there. (Actually there is a universe where this book exists, and the answer to every question is *lol nothing matters.*) But Robert Proctor, the professor of agnotology at Stanford, put it to me more delicately: "It's easy to exaggerate our importance in the universe."

Proctor likes to ask his students why they fear death and yet they don't fear the time before they were born. "There's a perfect temporal symmetry there—but no one fears the year 1215," as he put it.

Though it would be terrifying to live in 1215, his point is that we don't inherently fear a world without ourselves in it. What we fear is the world going *on* without us.

Ignorance of the dying process makes it all the more frightening, and this is another place where ignorance tends to be willful and self-imposed. There is much that can be known about the death of our bodies that is consoling and practical to consider. A metaphysical perspective begins concretely. For physicists to arrive at an idea as insane as a *multiverse,* they had to start by considering the most basic laws on tiny scales. That same approach can be used to appreciate our own place in the universe (multiverse), and the transient nature of our bodies. Dying can be viewed less as a cosmic injustice than an example of harmony and order on a larger scale.

So, to begin concretely: Almost immediately after death, a corpse begins teeming with bacterial life. We are largely composed of bacteria when we are alive—we are covered with them the moment we enter the world—but once we are corpses, look out. The proliferation of the "necrobiome," the colonies that largely arise from the microbiome, changing and proliferating based on the new terrain of a nonliving body, accounts for the hasty onset of the "dead body smell." It also accounts for the rapid disposal and disappearance of corpses when left in nature.

Yet the locked-in look and fetid odor of unadorned corpses are not the perception of death that most Americans have come to know. More than any other country in the world, by a wide margin, Americans practice open-casket ceremonies. Funeral homes will not agree to an open-casket ceremony unless the corpse has been thoroughly renovated. That process involves embalming, and its popularity neatly encapsulates much of the misunderstanding and ignorance with which Americans tend to think about death.

Many consider the fate of one's abandoned body taboo conversation beyond the two deeply ingrained options: embalmed or cremated? That is the narrow-minded view created by the multibillion-dollar "death industry," as marketing professor Susan Dobscha calls it. A student of consumer behavior, Dobscha became fascinated by the death industry in the middle of her career, when the partner of a close friend died. The two had been in a committed relationship for sixteen years, but unmarried because it was at the time illegal. So the funeral home would not release the body to his partner, who also felt pushback from funeral homes that would not conduct a "gay funeral."

But most fascinating to Dobscha was learning how the crematorium handled her friend's request to accompany the body. The man is from France, where it is standard for loved ones to go with the body to the crematorium. They don't witness the cremation, but they're there. In the United States this is completely unheard of. The crematorium insisted that he not come. It piqued Dobscha's interest in the consumer-behavior side of the death industry—an enormous field with which everyone deals but few speak frankly.

The median cost of embalming, a funeral service, and burial in the United States is $8,508. Some banks offer funeral loans. "Losing a loved one is hard enough without worrying about the cost of cremation or burial," offers First Franklin Financial, whose "funeral loans help relieve your financial stress so you can focus on what really matters."

What if what really matters is that people not take on additional debt in order to pay respectful homage to a person's life? Can such respect be paid without a velvet-lined casket and a formaldehyde-filled body?

The funeral industry has done well perpetuating the grand casket as a status symbol, one of love and respect for the deceased; but it is not part of any major religion, or of any cultural tradition that predates the more recent rise of the industry.

The burial "vault" that must be fitted to the outside of each casket costs, alone, an average of $1,327, according to the National Funeral Directors Association. Caskets cost much more. Even

Walmart, bastion of value, sells a mahogany casket for $3,499. That model is part of the upper cost echelon that is "sanctioned by the Vatican Observatory Foundation." A pared-down option, like the Star Legacy Natural Opulent Casket ($2,299) does not bear this approval. The Vatican Coffin is "a magnificent tribute to honor your loved one and celebrate a life of faith. Adorned with a high-gloss finish, premium swing bar handles, and a luxurious velvet interior, each casket is handcrafted and comes with a Certificate of Authenticity." (That assumes you might be resourceful enough to produce a counterfeit velvet-mahogany coffin, but not a passable Certificate of Authenticity? I don't know who goes to hell in that situation.)

Some coffins are advertised as air- and watertight, which seems like an upgrade, until you read about casket explosion. Colloquially known as "Exploding Casket Syndrome," it happens when bacteria metabolize a corpse, releasing gases inside a sealed casket. That increases pressure, effectively creating a bomb.

Even though it's rare for anyone to suggest that they would like to have their body buried at great expense financially and environmentally, and then potentially exploded, the continued practice is the result of failure to question the death industry's messages. As Dobscha explains it, our unthinking habits are driven by the casket industry, the funeral industry, and the embalming industry. "What they've done is picked up these functional products and made them about aesthetics and status."

Then there's the embalming process, which involves a mortician massaging rigid muscles until the limbs are pliable enough to be positioned in a lifelike fashion. Sometimes tendons must be cut. Caps are placed below the eyelids to make sure they remain fully closed throughout the viewing of the body. If the eyelids insist on opening, glue can be applied. Because the glands of the skin are no longer secreting oils, a mortician must apply copious creams. Cotton placed in the throat helps prevent embalming fluid from draining out through the mouth and nose. Cotton placed in the anus and vagina prevent embalming fluid "seepage"—a word that seems to appear in almost every sentence of the Funeral Consumers Alliance description of the embalming process. The mortician also cuts into a large vein, such as the femoral vein in the groin, and drains all of the blood from the body, then accesses an artery with a large-bore needle and fills all of the formerly blood vessels with several liters of embalming fluid. She cuts a hole in the belly button and inserts a tube through the muscular wall of the stomach and colon, attaching a vacuum that sucks out the contents of the entire gastrointestinal tract. By the same method, the lungs are collapsed and the chest cavity sucked dry. She then fills the chest and abdominal cavities with an especially concentrated embalming fluid. At this point the corpse is a liquid-filled shell. It is very heavy. It is washed. Its hair is combed. It is painted with cosmetics and manipulated into formalwear.

Two of my grandparents' bodies were embalmed, and at the time I chose not to think about the process. Dobscha felt more strongly. She remembers thinking that her "grandmother looked like a clown, being buried in makeup. She never wore it when she was alive. She would have been horrified." Muslims and Orthodox Jews consider embalming a desecration of the body; Christians are generally cool with it. Some even mistakenly believe it has a basis in their religious scripture or tradition. Islam has a dictum on burial, but it's antithetical to the kind of opulence of the conventional Western approach: The body is wrapped in a shroud and placed in a simple pine box and buried within forty-eight hours. This is, in Dobscha's findings, the most sustainable burial practice of any major religion. Decadence and embalming are nowhere prescribed in any Christian text or creed; they are merely a new-prosperity American tradition.

It is justified not by religion, then, but by a belief in the innate sanctity of any and all dead bodies. To many this means not just burying a corpse in a velvet-lined tomb, but doting over it, maybe hugging or kissing it. But the lengths we go to afford corpses the same respect as living people are deeply at odds with the grisly procedures required to prepare them for viewing.

"In the U.S., we are the most disconnected from the dead body of any culture," said Dobscha, who believes this is conflated with the American tradition of spending 80 percent of our health care expenses on people's last year of life keeping dying people alive. "That disconnection from the dead body gets to the larger issue that we are very disconnected from waste in general. Our food waste, our urine—all the stuff that comes out of our body."

Embalming became tradition in the United States after the Civil War rendered thousands of corpses in need of some preservation so that they could be transported for burial at home, when possible. A young physician named Thomas Holmes took to the task. Experimenting with various preservative chemicals to replace their blood, he was eventually able to embalm more than four thousand fallen Americans by his own count. This would be enough to earn him the moniker the "father of modern embalming." (*Modern* in this case draws a distinction between chemical preservation and the ancient practice of mummification.)

Five years after the end of the war, the German chemist August Wilhelm von Hofmann isolated the organic compound formaldehyde, which was extremely effective at preserving human tissue. The embalming fluid that is injected into corpses today is a petrochemical mixture based in formaldehyde that is banned in many countries. Unfortunately, formaldehyde was not classified as a carcinogen by the World Health Organization until the 1990s, only after a few generations of American corpses had been stuffed full of it and buried in the ground.

The adage "the dose makes the poison" is as critical with formaldehyde as it is with overdosing on water. Our bodies produce formaldehyde and use it to make amino acids, which we would die without. At that level, it is not a toxin. But as a biology lab at the National University of Singapore states in its procedural techniques for preserving fish specimens, using even just a diluted mixture of 96 percent water and 4 percent formaldehyde, "This is undoubtedly extremely painful for the animal, but death is usually fairly quick." (Scientists should receive more training in the use of adverbs.) Drinking just one ounce of formaldehyde will turn a living human, too, into a corpse, overwhelming the capacity of our enzymes to excrete it.

Beyond its toxicity from ingestion, formaldehyde has proven to cause nose cancer in humans and is strongly linked to leukemia and asthma, as well as spontaneous abortions. While performing embalming, morticians are required by federal law to wear full-body suits and respirators. Still, embalmers have high rates of leukemia and brain tumors, which appear to be due to the exposure.

To make matters worse, in 2014, ABC News reported, "Kids Use Embalming Fluid as Drug." I'm no expert on The Kids These Days, but I believe the same headline could be repurposed with almost any word in place of "embalming fluid." In this case, though, The Kids dip cigarettes into embalming fluid and smoke them. This is sometimes called "wet" and can also refer to mixing marijuana or tobacco with embalming fluid and/or PCP—which, confusingly, has also been known colloquially since the 1970s as "embalming fluid." Wet is actually not new at all, and has gone by various names, like water, fry, illy, and wack. For context, a user on one cannabis forum asked, "Where can i buy embalming fluid? Me and a friend wanted to try this so called 'wack' that when put on weed gets you so high you apparently can't even talk."

Psychopharmacologist Julie Holland attempted to distill

their motivation: "Embalming fluid appeals to people's morbid curiosity about death," she told ABC. "There's a certain gothic appeal to it."

There just have to be healthier ways to relate to death.

Still, the vast majority of embalming fluid is not smoked, but introduced to our ecosystems more quietly, buried in the ground. It is buried inside of millions of acres' worth of woodlands that we destroy to make opulent caskets. This is in addition to the acre of rainforest that we destroy every minute of every day in order to clear land for crops that are grown to feed livestock, to feed the ever-growing human population, thus hastening climate change, which is poised to cause famine and war, and more death. Trees are critical to reducing carbon dioxide that we pour into the air. They can't keep up, but their mitigating effect is not nothing. Yet we kill them to encase bodies that are already dead.

. . . Is there anything productive I can do with my dead body?

At Piedmont Pine Coffins in Bear Creek, North Carolina, Don Byrne is happy to fashion a custom coffin that's maximally biodegradable for a few hundred dollars. The unadorned box is amenable to personal touches—loved ones writing messages and children leaving handprints in paint. If, akin to many people's reasoning for an open-casket affair, a ritual is desired for closure or connection, he offers the opportunity for the family of the deceased to drive the final few screws into the coffin. (It's better to use screws in a thin-walled coffin than, as the adage goes, to put nails in the coffin, unless you mean to split it open.)

Or for $4.95, Byrne will email you his "DIY Plywood coffin plans." Others share similar instructions for free (including Chuck Lakin and company at lastthing.net, who kindly let us recreate them here).

If you need time to build the box and plan the funeral,

corpses can be preserved without embalming for many days in most hospital morgues' refrigeration units. "If you are a novice woodworker, take courage," Byrne implores on his website, assuring armchair craftspeople not only that the project can be completed in four hours for less than $200, but that "the satisfaction of your hands-on contribution to the undertaking of a loved one's funeral—or your own—is of great value."

If that value is $201 or more, you could actually profit from the endeavor. All that is needed is a screwdriver, a saw, a tape measure, a pencil, some plywood, a couple two-by-fours, and some two-inch screws. No power tools are necessary, a fact important to Byrne, who runs Piedmont Pine Coffins from his ranch, where he lives with no electricity or plumbing—only a pair of twelve-by-twelve-foot buildings (he spends most of his time outdoors), a garden, and a few animals. In a place like that, a buried body is pure boon to the ecosystem.

Cremation is a waste of fertilizer, though advantageous in that it does not require a burial plot. There are new sentimental ways to use ashes, like putting them into 3-D printers and creating customized vinyl records with them. Another company will mix ashes into concrete and drop them into the ocean to become part of a coral reef. The coral polyps adhere to the concrete-ash mixture the same way they adhere to other coral polyps. People can draw on the concrete and press their hands into it to decorate it prior to its submersion. The family is given a GPS coordinate where the person's ashes remain.

Minimal as cremation's impact is relative to embalming, every corpse that is burned emits inviolate fumes. Short of a sky burial—where a body is left atop a mountain to be picked apart by birds of prey—the most minimalistic option for many people might be a new one that is as yet illegal in many places.

In a process known as alkaline hydrolysis, or "green cremation," bodies are dissolved into a liquid. A strong solution of alkaline water (water with potassium hydroxide) disintegrates the body over the course of twelve hours. Heat and pressure can accelerate the process, which is carried out in a steel cylinder. According to Philip Olson, a professor of science and technology in society at Virginia Tech, "the effluent" can be disposed of through municipal sewer systems, provided the fluid is cooled

HOW TO BUILD

MATERIALS YOU'LL NEED:

2 SHEETS ½" × 4' × 8' PLYWOOD

3 OR 4 2" × 2" × 8'

1 LB SCREWS, 1¼"–1¾" LONG

12'–16' ROPE

STEP 1:

CUT THE PLYWOOD. CUT BOTH SHEETS THE SAME.

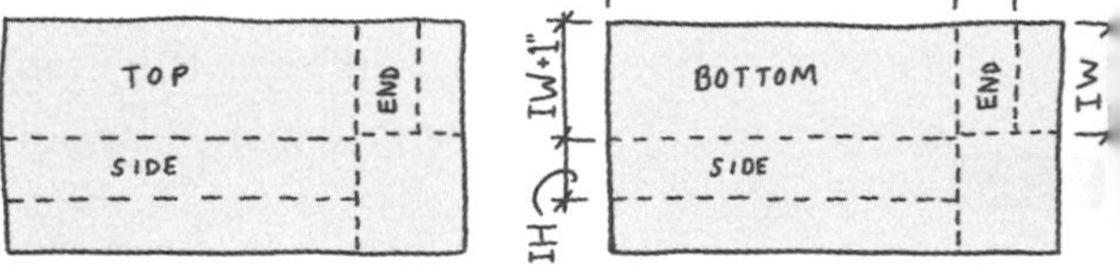

STEP 2:

DRILL THE HOLES. ⅛" FOR THE SCREWS, ½" FOR THE ROPE.

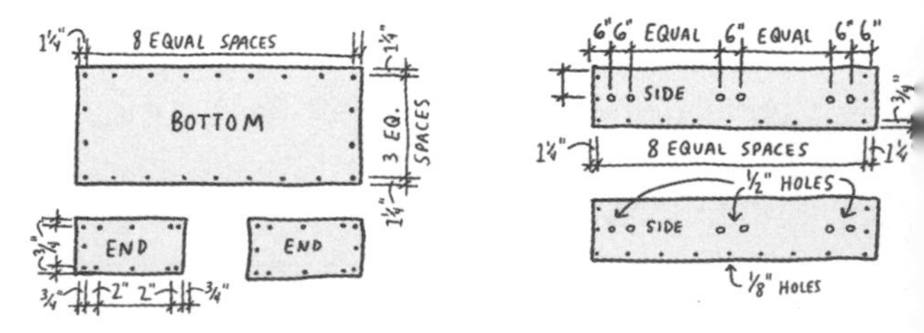

and the pH adjusted to meet local requirements. It uses much less energy than cremation but more water—three hundred gallons per corpse. Long used in research facilities to dispose of tissue specimens, only recently has the process been expanded to dispose of entire human bodies.

In 2011, Jeff Edwards became the first funeral director to offer

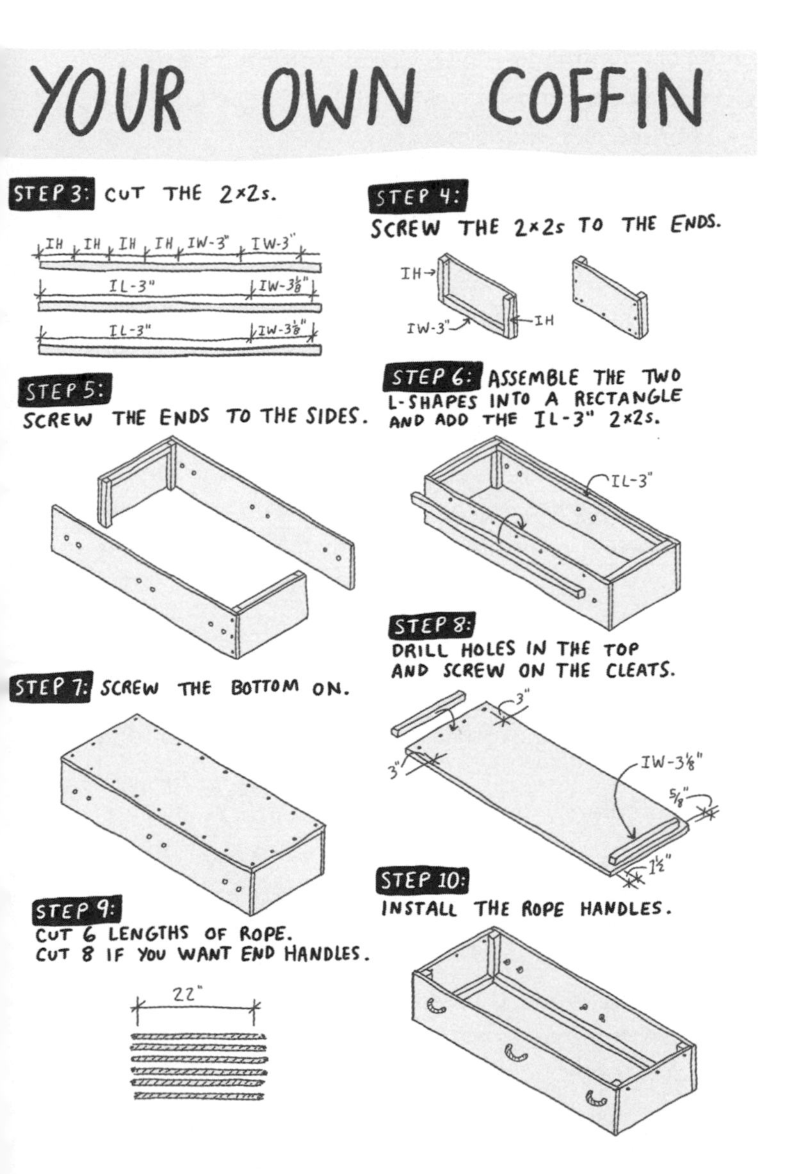

alkaline hydrolysis. He hydrolyzed just nineteen corpses before the Health Department of Ohio said that it would not issue death certificates for corpses that had been so disposed. Though, as he saw it, the concerns were not really about health. Edwards accused the department of "buckl[ing] to the intimidation or threats of some of my local competitors in the funeral business."

This influence may be the reason that the practice is yet legal in only nine U.S. states and one Canadian province. It is a threat to the death status quo that is eerily similar to the mid-twentieth-century rise of cremation.

For decades, funeral directors portrayed cremation as "undignified, irreligious, and even un-American," Olson recounts. The smear campaign ended abruptly, though, when the industry finally decided to capitalize on the demand. It takes time for any industry to adapt to disruptive technology, but especially when the industry is built on maintaining arbitrary perceptions—in this case, an extreme reverence for dead bodies.

"Adversaries of alkaline hydrolysis commonly express disgust at the thought of sacred human remains being flushed down the drain like everyday bodily waste, and at the thought of those remains somehow finding their way back into the bodies of the living," writes Olson. It is the same principle of circularity in life that to others provides closure. When I asked Susan Dobscha about the idea of closure in an open casket and ceremonially embalmed body, she countered that there was more closure in the idea of a body going back to the earth and being fuel for a tree.

In the spirit of circularity, too, there is always the standby of "donating your body to science." Anyone who has ever said, 'I hope they find a cure for [some malady]' is disingenuous if they then proceed to have their corpse filled with formaldehyde and buried in the ground.

Like everyone who went through medical school in the United States, I dissected a cadaver. It was a respectful and clinical environment, and the educational value was unparalleled.

But that's just one of the many valuable uses of a body donated to science. Soon it could be just as valuable to donate only a few cells.

What happens to the data version of me when I die?

For years, people exiting Gunther von Hagens's Body Worlds were invited to fill out a form that indicated they wished to one day donate their own bodies to the exhibit.

The eccentric doctor's display, consequently, stood apart from the nearly identical but totally unrelated exhibit called BODIES. While Body Worlds insists that all corpses are procured from people who consent to their use, specifically in that exhibit, BODIES—which came onto the scene ten years after Body Worlds, using the same plastination technique invented by Von Hagens—openly admits that they *do not* use willing donors, but exclusively Chinese bodies of unknown origin.

If there are ethical dilemmas around the practice of paying donors of blood and organs—because of the black market that it creates, and the coercion implicit in such a system—a black market for corpses is only more vexing. But, somehow, BODIES is still thriving after more than a decade and fifteen million paying customers. I only recently passed up an opportunity to see it at the Luxor hotel and casino in Las Vegas, where the corpses-of-unknown-origin from a country with a shoddy human rights record represent part of an "entertainment" lineup that includes the illusionist Criss Angel and the tragedy Carrot Top.

Domain over our bodies is among the most complex and consequential questions before medical science, and rarely is the answer so straightforward as in the case of a body sold and displayed without consent.

Most governments today do acknowledge that as long as the tissues that compose your organs are still physically attached to you, you own and control those cells. Which sounds like good news, except that if you've ever had a Pap smear or a drug test or a skin biopsy or a haircut, you know how easy it is to surrender tissues—which can be used by anyone for anything without consent.

Still, few people realize that this can happen, and that we have no rights to innovation that comes as a result of the work done on our tissues. The most famous example of this is the case of Henrietta Lacks, an impoverished black woman who died of cervical cancer in 1951. Doctors at Johns Hopkins took a biopsy of her cervix and grew the telomerase-producing (immortal) cancer cells in a culture that eventually grew to become trillions of cells, sold to researchers around the world. Lacks's cells have lived longer outside of her body than they did inside it. Her family gave no permission and received no compensation for the role of her cells in thousands of patents and important advancements, which ranged from the polio vaccine to the basis of in vitro fertilization to the understanding that humans have forty-six chromosomes.

In the United States, the provision that cedes control of a person's tissues upon detachment from their bodies is called the Common Rule. Drafted in 1979, it was intended to ensure the free enterprise of science, but it has proven less straightforward as technology has advanced.

The potential for what can be done with our tissues will become the central ethical question of biomedical science driven by advances in mapping DNA that have happened at rates few imagined. Moore's law for computers says that capacity goes up every eighteen months by a factor of two, and cost comes down by the same amount. The process of sequencing human DNA has blown that away, both in terms of increasing throughput and decreasing cost. When Francis Collins and colleagues at the U.S. National Institutes of Health sequenced the first human genome, it cost around $400 million. Now, thirteen years later, the cost is $1,000.

"I don't think we're hitting any law of physics here," Collins told me. "I think we're going to continue to see this drop. I would guess we'll get to the $100 genome in the next decade."

At that point, I should update this book to include a section called "The Data Parts." But in the immediate term, there are questions that everyone would do well to start thinking about.

Anticipating this landslide of bodily information, there is already consensus among scientists that people must retain the right to determine the fate of their own bodily data, their genomes and medical records and fitness and dietary logs, and soon their microbiomes, and metabolomes, exposomes, proteomes, and potentially other bodily processes that can be quantified into *omes*. The question is how exactly all of this information about our bodies can be managed responsibly, safely, and productively.

Today it is illegal in the United States to experiment on a person without consent. And though the Common Rule provision has led to discoveries that have saved many lives, it is based on ethical principles of four decades ago. The double helix had only just made its way into textbooks; the human genome would not be sequenced for three decades. The Common Rule could not have anticipated that "anonymous" tissue samples could be put through DNA sequencing and then traced back to people. In the worst hypothetical scenarios, this information could be used to discriminate against those attempting to purchase insurance or procure a mortgage, or it could be used to frame a person for murder.

Outlandish as those scenarios may sound to some, to others they are totally plausible. Starting in the 1930s, under the "Tuskegee Study of Untreated Syphilis in the Negro Male," federally funded researchers caused hundreds of mostly poor, illiterate sharecroppers to suffer and die of syphilis. The subjects died so that researchers could watch and learn how an infection of syphilis spirochetes spreads from the genitals into the brain and spine; how it paralyzes and blinds a person and causes madness.

The highly effective treatment penicillin had been available since 1947, yet these unwilling farmers continued to die until journalist Jean Heller broke the story in 1972.

Laws guaranteeing creative scientific liberty with this data exacerbate distrust in the medical establishment, which the Lacks family, like so many black Americans, feared was more interested in using them as test subjects than in caring for them.

To begin to address this, in 2016, the U.S. Department of Health and Human Services suggested that people should have to sign consent forms before anything is done to their tissues. Critical as it may be that researchers have access to enormous pools of data and cell lines available for creative, open-ended global collaboration, asking people to sign a similarly open-ended consent statement is a request for faith that does not everywhere exist. A consent form could at best read something like, *We want to use your tissue for research, though that could mean anything. Cool?*

Still, many brilliant people tell me they would sign. It was when I was in Vegas at a health-tech conference (and not seeing BODIES) that I first met the physician and Harvard Medical School professor John Halamka. We bonded over the difficulty procuring plant-based food in that city. (So, basically it was an outtake from *The Hangover*.) Halamka was the second person to have his genome sequenced. He uploaded all of it to the Internet. He obsessively tracks his life with his Fitbit and his iPhone and his mind, and he uploads all records to the cloud just in case—sooner or later—any of it can be useful to advance medical research in any way. There are nine petabytes of John Halamka on the Internet, if you want them. You can even get some of his stem cells, for a modest honorarium.

Halamka's decision can be understood only in the context of historical trust and distrust of health care. He is a white male vegan MD/PhD who lives on a farm in the Boston suburbs. He is a polymath who makes apps and essentially cured his wife's breast cancer by analyzing her genes. His decision to share the data of his life is deeply informed and purely elective. His situation could not be further from that of Henrietta Lacks.

Not only is he unconcerned about people abusing the data from his body, but he is even more unusual among scientists in that he seems unconcerned with people profiting from it.

In an ideal world, if someone uses your idea, or data you worked to acquire, and uses it to make a major advance in medicine, then everyone is happy. In the real world, there is animosity and resentment over who gets credit, who profits, who gets job security and prestige and funding for future projects. In this way, hoarding of data has slowed progress.

Accordingly, Francis Collins at the National Institutes of Health is attempting to teach scientists to share. "People say it's not going to work, because it's like herding cats," he told me. "Well, you know, you can't herd cats. But you can move their food. We at NIH have a lot of food for the cats," he said, referring to the $32 billion in taxpayer money that his institution distributes to scientists every year. "And we're aiming to use it in that way—to really provide incentives for people to open up their data access."

During the launch of the federal Precision Medicine Initiative in 2016—which aims to collect a million human genomes into a public database—the director of the White House office of Science and Technology Policy, John Holdren, tried to assure me that the top priority will be security. The U.S. government alone will soon be managing a cloud of human biological data that is exobytes in size—billions of gigabytes.

Of course, if the *top* priority were security, we would never acquire any data. We would never put anything on the Internet, and never get blood tests or leave a fiber of our hair anywhere, which means never do anything. As in all of life, there are risks and benefits in choosing to act, and in choosing not to.

Making peace with all of this, with the new mortality, seems to involve keeping sight of Franklin's appraisal of our inconsequential place in the universe, as well as Halamka's approach to what can be done to improve humanity. He has achieved immortality on his own terms, in a way that does not fill the planet with thousand-year-olds. What will remain of him when he dies will exist in a cloud, to help the people who still have bodies make the most of their time.

EPILOGUE

I appreciate your sticking with me through these stories, which, I realize, are in most cases less straightforward as answers than might seem ideal. We're not that straightforward, as a species. So you are just as much to blame as I am. I went to medical school expecting to come away with some sort of mastery of how we work, and I ended up with only more questions. The more I learn, the more questions I have.

That could leave me feeling nihilistic, and sometimes it has, but more often it's been beneficial to my sense of empathy. Taking time to embrace our complexity is, I think, more useful and empowering than memorizing trivia-style explanations. The decisions we face about our bodies demand that we constantly challenge definitions of normalcy and health, that we consider how we want to live and die in advance of major crossroads, and that we think about the effects of our decisions on the greater human body as a species within an ecosystem.

At the Spotlight Health conference in Aspen in the summer of 2016, I got to listen to a panel of eminent scientists, including Nobel laureate David Baltimore, who have been at the heart of the long-emerging field of gene editing. They explained how to alter human germ cells to create a disease-free human (in principle, as in, a person guaranteed to be born without sickle-cell anemia, despite two parents who carry the gene). In the audience was the former head of the FDA, Frank Young, who said afterward that he was at a loss for how such technology would be regarded by society. Is it a drug? Should anyone have access to it? Will people abuse it? Are there reasons not to continue pursuing it?

These are questions that will affect not just the ecosystems of our own bodies, but the systems that extend to the planet and the people with whom we share it. These big questions about gene editing and data sharing all felt pretty abstract to me until I met the Kopelans. Rafi's dad, Brett, still cautiously optimistic about a cure for his daughter's DEB, bursts when he talks about the possibilities. "It's amazing what can happen. Is a PhD in biology going to solve a disease, or in bioinformatics? Could be a mathematician."

John Holdren, who has led the government's initiative to pool our data, believes that "humanity-minded" people will donate their information to the cause, as people have long donated their organs and bodies. As he put it to me, "Everybody has relatives who are suffering from a wide range of diseases, and everybody knows that ultimately they will get sick of something. That personal reality is a powerful force."

It is what drives every quotidian decision—about what to eat and drink, how to modify ourselves outwardly and inwardly, whether to call the police when we see nipples, who to have sex with and how, who to keep close, who to push away, and how to feel about all of those decisions. We can regard ourselves as Max Factor would have us do, calculating and quantifying our deficiencies and attempting to make them up, ad infinitum until death, or we can abandon arbitrary standards in favor of a fluid world.

Technology is propelling medical science so quickly that much of what doctors do in practice today was not taught to them in medical school. Gastroenterologists (and neurologists and dermatologists and dieticians) are now reckoning with our trillions of microbes that are clearly relevant to almost everything our bodies do, and yet were largely unknown to us even a few years ago.

As doctors and patients, the best we can do is prepare ourselves to learn about the next things as they come, as they inevitably will, in the gradual, continuous process. I hope that this book is helpful in contextualizing medical knowledge in a way that minimizes undue concern, or at least helps you to prioritize concerns for yourself. I also hope it's clear by now that this guide to operating and maintaining a human body is no guide in any prescriptive sense. It's rather about maximizing autonomy, guiding only in a fundamental way: to encourage questioning of the cultural and commercial messages all around us, challenging normalcy, and remaining skeptical of simplistic solutions.

Unless there's a contact lens in your brain.

ACKNOWLEDGMENTS

To the readers who keep up with what I do and, so, made this possible. To all who offered questions to be answered here, whether the questions made the final manuscript or not. To the sources who shared their stories and ideas. To all the people whose writing and speaking I've drawn from. Though I've worked to give credit in these notes and bibliography and within the text, I also felt it impossible to name everything I've drawn on from years of reading, watching, interviewing, and listening. In an attempt to make this book comprehensive, I reached well outside of my personal experience and expertise—to such a degree that could warrant a footnote at the end of most sentences. Conversations and secondary sources led me to primary sources that went uncited in the milieu, and yet without whom this could not have existed. Thank you to all of those writers and thinkers for their work. Especially to the journalists whose stories I've been inspired by and mentioned here, I'm grateful and hope this book serves to bring only more attention to your work and the causes you further, and to the importance of the institution of which I'm privileged to be a part.

For editing this book, Yaniv Soha. For fact-checking, Julie Tate. For illustrations, Hallie Bateman.

For their work in bringing the video series "If Our Bodies Could Talk" to life: Katherine Wells, Nic Pollock, Jaclyn Skurie, David Sidorov, Nadine Ajaka, Jackie Lay, Matt Ford, Chris Heller, Paul Rosenfeld, Sam Price-Waldman, Jake Swearingen, and Kasia Cieplak-Mayr von Baldegg. As well as friends who've helped with the show: David Young, Chiara Atik, Amy Rose Spiegel, Mark Bittman, and all other guests who have given their time.

For giving me the opportunity to be a part of this generation of *Atlantic* editors: John Gould, Bob Cohn, James Bennet, Scott Stossel, Alexis Madrigal, Derek Thompson, Hayley Romer, David Bradley.

And for support in the process and along the way: Mom, Dad, Sarah Yager, Lauren Hamblin, R. James Hamblin, Eleanor Hamblin, Richard

Johns, Norma Johns, Avi Gilbert, Eric Lupfer, Margo Shickmanter, Sarah Porter, Michael Goldsmith, Ross Andersen, Dan Buettner, Melissa Price, Rob Su, Steve Futterer, Bob Ryu, Richard Gunderman, Jody Avirgan, Maeve Higgins, Julie Beck, Lex Berko, Lindsay Abrams, Chris Cillizza, Lizzie O'Leary, Spencer Kornhaber, Rebecca Greenfield, Mallory Ortberg, Alex Bellows, Lucie Coneys, Cameron Cox, Anna Bross, Jeff Goldberg, Alexandra Chabrerie, and many others.

NOTES

PROLOGUE

xiii **There are common:** Yih-Chung Tham et al., "Global Prevalence of Glaucoma Projections of Glaucoma Burden Through 2040," *American Association of Ophthalmology Journal* 121, no. 11 (November 2014): 2081–90.

xv **"In one oft-repeated pattern":** Dobscha, Susan, *Death in a Consumer Culture* (New York: Routledge, 2015), 251.

xx **He calls this instinct:** Paul Rozin, Michele Ashmore, and Maureen Markwith, "Lay American Conceptions of Nutrition: Dose Insensitivity, Categorical Thinking, Contagion, and the Monotonic Mind," *Health Psychology* 15, no. 6 (November 1996): 438–47, doi:http://dx.doi.org/10.1037/0278-6133.15.6.438.

xxii **Illustration:** Thomas R. Frieden, "A Framework for Public Health Action: The Health Impact Pyramid," *American Journal of Public Health* 100, no. 4 (2010): 590–95, accessed September 22, 2016, https://www.ncbi.nlm.nih.gov/pmc/articles/PMC2836340/.

xxiv **Illustration:** Steven A. Schroeder, "We Can Do Better—Improving the Health of the American People," *New England Journal of Medicine* 357 (2007): 1221–28, accessed September 22, 2016, http://www.nejm.org/doi/full/10.1056/NEJMsa073350.

ONE • APPEARING: The Superficial Parts

5 **An elaborate hood of metallic bands:** Adrienne Crezo, "Dimple Machines, Glamour Bonnets, and Pinpointed Flaw Detection," *The Atlantic,* October 3, 2012.

7 **"The thing that moves us":** Thomas J. Scheff, "Looking Glass Selves: The Cooley/Goffman Conjecture," August 2003, www.soc.ucsb.edu/faculty/scheff/19a.pdf.

8 **It's an anatomical anomaly:** "Genetic Traits: Dimples," Genetic Index, www.genetic.com.au/genetic-traits-dimples.html.

9 **In 1936, entrepreneur:** "Woman Invents Dimple Machine," *Mod-*

ern Mechanix, October 1936, http://blog.modernmechanix.com/woman-invents-dimple-machine/.

13 **"In some cases":** Morad Tavallali, "Cheek Dimples," Tavallali Plastic Surgery, www.tavmd.com/2012/06/30/cheek-dimples/.

13 **As a spokesperson:** "British Doctors Warn Against 'Designer Dimple' Cosmetic Surgery," *Herald Sun* (Australia), June 22, 2010, www.heraldsun.com.au//news/breaking-news/british-doctors-warn-against-designer-dimple-cosmetic-surgery/story-e6frf7jx-1225882980055.

15 **One of the most interesting statistics:** "Amazing Facts About Your Skin, Hair, and Nails," American Academy of Dermatology, 2016, www.aad.org/public/kids/amazing-facts.

16 **The "golden rule" in tattooing:** "Top 5 Reasons for Removing Tattoos," Fallen Ink Laser Tattoo Removal, www.falleninktattooremoval.com/2014/12/11/top-5-reasons-for-removing-tattoos/.

16 **Spending on tattoo removal:** Quentin Fottrell, "Even Before Apple Watch Snafu, Tattoo Removal Business Was Up 440%," MarketWatch, May 2, 2015, www.marketwatch.com/story/tattoo-removal-surges-440-over-the-last-decade-2014-07-15.

18 **In the journal *Nature:*** Katherine D. Zink and Daniel E. Lieberman, "Impact of Meat and Lower Palaeolithic Food Processing Techniques on Chewing in Humans," *Nature* 531 (March 24, 2016): 500–503, doi:10.1038/nature16990.

18 **Many anthropologists believe:** Aaron Blaisdell, Sudhindra Rao, and David Sloan Wilson, "How's Your Ancestral Health?," *This View of Life,* Evolution Institute, March 24, 2016, https://evolution-institute.org/article/hows-your-ancestral-health/.

23 **Photographs of Eli Thompson:** Salvador Hernandez, "Meet the Very Cute Baby Who Was Born Without a Nose," *BuzzFeed News,* March 31, 2015, www.buzzfeed.com/salvadorhernandez/meet-the-very-cute-baby-who-was-born-without-a-nose?utm_term=.kcWrVWNqJ#.dnWY8qorz.

25 **The nose actually begins:** Mao-mao Zhang et al., "Congenital Arhinia: A Rare Case," *American Journal of Case Reports* 15 (March 18, 2014): 115–18, doi:10.12659/AJCR.890072.

26 **The celebrity Elizabeth Taylor:** Soheila Rostami, "Distichiasis," Medscape, October 14, 2015, http://emedicine.medscape.com/article/1212908-overview.

26 **People with an extra row:** "Lymphedema-Distichiasis Syndrome," Genetics Home Reference, February 2014, https://ghr.nlm.nih.gov//condition/lymphedema-distichiasis-syndrome.

28 **After cataracts, glaucoma:** "Causes of Blindness and Visual Impairment," World Health Organization, www.who.int/blindness/causes/en/.

28 **And so they tested:** Guillermo J. Amador et al., "Eyelashes Divert

Airflow to Protect the Eye," *Journal of the Royal Society Interface*, February 25, 2015, http://rsif.royalsocietypublishing.org/content/12/105/20141294.

29 **"We combine precision":** J. T. Miller et al., "Shapes of a Suspended Curly Hair," *Physical Review Letters* 112, no. 6 (February 14, 2014), http://dx.doi.org/10.1103/PhysRevLett.112.068103.

31 **Illustration:** B. Wang et al., "Keratin: Structure, Mechanical Properties, Occurrence in Biological Organisms, and Efforts at Bioinspiration," *Progress in Materials Science* 76 (2016): 229–318, accessed September 22, 2016, http://dx.doi.org/10.1016/j.pmatsci.2015.06.001.

32 **Waxing, shaving:** "Amazing Facts About Your Skin, Hair, and Nails," American Academy of Dermatology, https://www.aad.org/public/kids/amazing-facts.

35 **His story is explained:** "How to Grow 3–6 Inches Taller in 90 Days," YouTube, October 12, 2012, http://tune.pk/video/4890970/how-to-grow-3-6-inches-taller-in-90-days-lance-story.

36 **In their dominant arms:** Hartmut Krahl et al., "Stimulation of Bone Growth Through Sports: A Radiologic Investigation of the Upper Extremities in Professional Tennis Players," *American Journal of Sports Medicine* 22, no. 6 (1994), doi:10.1177/036354659402200605.

37 **According to researcher:** Richard Knight, "Are North Koreans Really Three Inches Shorter Than South Koreans?," BBC News, April 23, 2012.

37 **Schwekendiek explains:** Ibid.

38 **The World Food Programme estimates:** "Hunger Statistics," www.wfp.org/hunger/stats.

38 **When the American astronaut:** Felix Gussone and Shelly Choo, "NASA's Scott Kelly Grew 2 Inches: The Body After a Year in Space," NBC News, March 3, 2016.

46 **Only two-thirds:** A. E. Davies et al., "Pharyngeal Sensation and Gag Reflex in Healthy Subjects," *Lancet* 345, no. 8948 (February 25, 1995): 487–88.

47 **Psychologists have shown:** "Low-Voiced Men Love 'Em and Leave 'Em, Yet Still Attract More Women: Study," EurekaAlert!, October 16, 2013, www.eurekalert.org/pub_releases/2013-10/mu-lml101613.php.

48 **When a professional baseball player:** Culley Carson III, "Testosterone Replacement Therapy for Management of Age-Related Male Hypogonadism," Medscape, 2007, www.medscape.org/viewarticle/557247.

50 **In the United States:** Cecilia Dhejne et al., "Long-Term Follow-Up of Transsexual Persons Undergoing Sex Reassignment Surgery: Cohort Study in Sweden," *PLoS ONE* 6, no. 2 (February 22, 2011), http://dx.doi.org/10.1371/journal.pone.0016885.

51 **In his 1964 State of the Union address:** Lyndon Baines Johnson,

"State of the Union Address," January 8, 1964, www.americanrhetoric.com/speeches/lbj1964stateoftheunion.htm.

53 **Instead of targeting:** Robert Rector and Rachel Sheffield, "The War on Poverty After 50 Years," Heritage Foundation, September 15, 2014, www.heritage.org/research/reports/2014/09/the-war-on-poverty-after-50-years.

53 **As St. John's continued:** "St. John's Well Child and Family Center," Southside Coalition of Community Health Centers, http://southsidecoalition.org/stjohns/.

54 **In the current accepted parlance:** "PFLAG National Glossary of Terms," PFLAG, www.pflag.org/glossary.

TWO • PERCEIVING: The Feeling Parts

62 **An international consortium:** S. M. Langan, "How Are Eczema 'Flares' Defined? A Systematic Review and Recommendation for Future Studies," *British Journal of Dermatology* 170, no. 3 (March 12, 2014): 548–56, doi:10.1111/bjd.12747.

62 **German professors once demonstrated:** V. Niemeier and U. Gieler, "Observations During Itch-Inducing Lecture," *Dermatology Psychosomatics* 1, no. 1 (June 1999): 15–18, doi:10.1159/000057993.

64 **Illustration:** Bill Gates, "The Deadliest Animal in the World," *Gates Notes* (blog), April 25, 2014, https://www.gatesnotes.com/Health/Most-Lethal-Animal-Mosquito-Week.

65 **Ramachandran's mirror approach:** Atul Gawande, "The Itch," *The New Yorker,* June 30, 2008, www.newyorker.com/magazine/2008/06/30/the-itch.

65 **It was from Washington University:** Yan-Gang Sun and Zhou-Feng Chen, "A Gastrin-Releasing Peptide Receptor Mediates the Itch Sensation in the Spinal Cord," *Nature* 448 (July 25, 2007): 700–703, doi:10.1038/nature06029.

66 **In one desperate case**: Marie McCullough, "Exploring Itching as a Disease," Philly.com, January 20, 2014, http://articles.philly.com/2014-01-20/news/46349734_1_itch-and-pain-pain-clinics-skin.

69 **Which it would be:** Matthew Herper, "Why Vitaminwater Is Bad for Public Health," *Forbes,* February 8, 2011.

70 **In 1933, ascorbic acid:** Kenneth J. Carpenter, "The Discovery of Vitamin C," *Annals of Nutrition and Metabolism* 61, no. 3 (November 26, 2012): 259–64, doi:10.1159/000343121.

74 **In 2016, Stevens:** Aswin Sekar et al., "Figure 5: C4 Structures, C4A Expression, and Schizophrenia Risk" (chart), in "Schizophrenia Risk from Complex Variation of Complement Component 4," *Nature* 530 (February 11, 2016): 177–83, doi:10.1038/nature16549.

74 **Also in 2016:** Soyon Hong et al., "Complement and Microglia Mediate

Early Synapse Loss in Alzheimer Mouse Models," *Science* 352, issue 6286 (March 31, 2016): 712–16, doi:10.1126/science.aad8373.

75 **"Health is cool":** "Health Myths Debunked—with Dave Asprey LIVE at the Longevity Now® Conference 2014," YouTube, May 23, 2014, www.youtube.com/watch?v=sHq_Xvu03zk.

77 **Caffeine can similarly improve:** David Venata et al., "Caffeine Improves Sprint-Distance Performance Among Division II Collegiate Swimmers," *Sport Journal,* April 25, 2014.

77 **Eighty-five percent of U.S. adults:** Diane C. Mitchell et al., "Beverage Caffeine Intakes in the U.S.," *Food and Chemical Toxicology* 63 (January 2014): 136–42, www.sciencedirect.com/science/article/pii/S0278691513007175.

77 **(At least it did):** Keumhan Noh et al., "Effects of Rutaecarpine on the Metabolism and Urinary Excretion of Caffeine in Rats," *Archives of Pharmacal Research* 34, no. 1 (January 2011): 119–25, doi:10.1007/s12272-011-0114-3.

78 **if people start filing grievances:** "Supplements and Safety," *Frontline,* PBS, January 2016.

82 **"it keeps the radiation":** Hochman, David, "*Playboy* Interview: Sanjay Gupta," *Playboy,* August 12, 2015, www.playboy.com/articles/playboy-interview-sanjay-gupta.

83 **On cue, coinciding:** Russell Brandom, "The New York Times' Smartwatch Cancer Article Is Bad, and They Should Feel Bad," *The Verge,* March 15, 2015, www.theverge.com/2015/3/18/8252087/cell-phones-cancer-risk-tumor-bilton-new-york-times.

83 **In 2016, he was punished:** Truman Lewis, "Feds Draw Blinds on Mercola Tanning Beds: The Company Claimed Indoor Tanning Was Safe, Did Not Cause Skin Cancer and Could Delay Aging," *Consumer Affairs,* April 14, 2016, www.consumeraffairs.com/feds-draw-blinds-on-mercola-tanning-beds-041416.html.

83 **Five years prior:** Steven Silverman, "Inspections, Compliance, Enforcement, and Criminal Investigations," U.S. Food and Drug Administration, March 22, 2011.

84 **Because these claims:** Ibid.

87 **In the brains:** Paul Thibodeau et al., "An Exploratory Investigation of Word Aversion," https://mindmodeling.org/cogsci2014/papers/276/paper276.pdf.

87 **Twenty percent of Americans:** Paul Thibodeau et al., "An Exploratory Investigation of Word Aversion," https://mindmodeling.org/cogsci2014/papers/276/paper276.pdf.

88 **A fifty-eight-year-old:** Mari Jones, "Tragic Dad 'Driven to Suicide Couldn't Face Another Day with the Unbearable Pain of Tinnitus,'" *Mirror* (UK), July 30, 2015.

88 **A London guitarist:** "Rock Music Fan 'Stabbed Himself to Death in

Despair' After Three Months of Tinnitus Made His Life Hell," *Daily Mail* (UK), November 19, 2011.

88 **In the Netherlands:** "Is Suicide the Only Cure for Tinnitus? It Was for Gaby Olthuis . . . ," StoptheRinging.org, March 20, 2015, www.stoptheringing.org/is-suicide-the-only-cure-for-tinnitus-it-was-for-gaby-olthuis/.

89 **Audiologist Allen Rohe:** Debbie Clason, "Tinnitus and Suicide: Why It's Happening, How to Stop It," Healthy Hearing, October 24, 2014, www.healthyhearing.com/report/52313-Tinnitus-and-suicide-why-it-s-happening-how-to-stop-it.

90 **The Office of Dietary Supplements:** Institute of Medicine, Food and Nutrition Board, *Dietary Reference Intakes for Vitamin A, Vitamin K, Arsenic, Boron, Chromium, Copper, Iodine, Iron, Manganese, Molybdenum, Nickel, Silicon, Vanadium, and Zinc* (Washington, DC: National Academies Press, 2001).

91 **Just a half cup:** Office of Dietary Supplements, National Institutes of Health, "Vitamin A: Fact Sheet for Health Professionals," February 11, 2016, https://ods.od.nih.gov/factsheets/VitaminA-HealthProfessional/.

91 **One 2015 study:** Tea Lallukka et al., "Sleep and Sickness Absence: A Nationally Representative Register-Based Follow-Up Study," *Sleep* (September 1, 2014): 1413–25, doi:10.5665/sleep.3986.

92 **Illustration:** Tea Lallukka et al., "Sleep and Sickness Absence: A Nationally Representative Register-Based Follow-Up Study," *Sleep* 37, no. 9 (2014): 1413–25, http://doi.org/10.5665/sleep.3986.

97 **In Indonesia:** Sanskrity Sinha, "Mita Diran, Indonesian Copywriter, Dies After Working for 30 Hours," *International Business Times,* December 19, 2013, www.ibtimes.co.uk/mita-diran-indonesian-copywriter-dies-after-working-30-hours-1429583.

98 **"There are many stories":** "Monster Energy Drink Deaths and Hospitalizations," LawyersandSettlements.com, October 19, 2015, www.lawyersandsettlements.com/lawsuit/monster-energy-drink-deaths-hospitalizations.html?opt=b&utm_expid=3607522-13.Y4u1ixZNSt6o8v_5N8VGVA.1&utm_referrer=https%3A%2F%2Fwww.google.com.

100 **In one study published:** H. P. Van Dongen et al., "The Cumulative Cost of Additional Wakefulness: Dose-Response Effects on Neurobehavioral Functions and Sleep Physiology from Chronic Sleep Restriction and Total Sleep Deprivation," *Sleep* 26, no. 2 (March 15, 2003): 117–26, http://www.ncbi.nlm.nih.gov/pubmed/12683469.

112 **Hemispherectomy for severe epilepsy:** Michael S. Duchowny, "Hemispherectomy and Epileptic Encephalopathy," *Epilepsy Currents* 4, no. 6 (2004): 233–35, doi:10.1111/j.1535-7597.2004.46007.x.

113 **For many people:** Seth Wohlberg and Debra Wohlberg, "www.gracewohlberg.blogspot.com: January 2009 to March 2010," http://www.rechildrens.org/images/stories/Graceblog.pdf.

THREE • EATING: The Sustaining Parts

126 **Cobain came to consider:** Lynn Cinnamon, "Cobain's Disease & Kurt's Sick Guts," *Lynn Cinnamon* (blog), April 22, 2015, http://lynncinnamon.com/2015/04/cobains-disease-kurt-cobains-sick-guts/; "Kurt Cobain Talks Music Videos, His Stomach & Frances Bean | MTV News," YouTube, www.youtube.com/watch?v=hJtm9HomKdE.

127 **Only in a 2011 article:** Emeran A. Mayer, "Gut Feelings: The Emerging Biology of Gut-Brain Communication," *Nature Reviews Neuroscience* 12, no. 8 (2011): 453–66, doi:10.1038/nrn3071.

130 **In 2010, a British woman:** Abhishek Sharma et al., "Intractable Positional Borborygmi—an Unusual Cause Diagnosed by Barium Contrast Study," *BMJ Case Reports* 2010 (2010), doi:10.1136/bcr.01.2010.2637.

132 **sleep-deprived subjects gained:** A. M. Spaeth, "Effects of Experimental Sleep Restriction on Weight Gain, Caloric Intake, and Meal Timing in Healthy Adults," *Sleep* 36 (7): 981–90, www.ncbi.nlm.nih.gov/pubmed/23814334.

134 **The people called it *beriberi:*** J. Ridley, "An Account of an Endemic Disease of Ceylon, entitled Berri Berri," in James Johnson, *The Influence of Tropical Climates on European Constitutions* (London: Thomas and James Underwood, 1827).

134 **He was puzzled:** Kenneth J. Carpenter, "Studies in the Colonies: A Dutchman's Chickens, 1803–1896," chap. 3 in *Beriberi, White Rice, and Vitamin B: A Disease, a Cause, and a Cure* (Berkeley: University of California Press, 2000), 26.

138 **After twenty-nine people filed:** U.S. Food and Drug Administration, "FDA Warns Consumers About Health Risks with Healthy Life Chemistry Dietary Supplement: Laboratory Tests Indicate Presence of Anabolic Steroids," July 26, 2013.

138 **When the FDA threatened:** U.S. Food and Drug Administration, "Purity First Health Products, Inc. Issues Nationwide Recall of Specific Lots of Healthy Life Chemistry B-50, Multi-Mineral and Vitamin C Products: Due to a Potential Health Risk," July 31, 2013.

138 **more than a decade of anabolic steroid use:** "Hulk Hogan, on Witness Stand, Tells of Steroid Use in Wrestling," *The New York Times,* July 15, 1994, www.nytimes.com/1994/07/15/nyregion/hulk-hogan-on-witness-stand-tells-of-steroid-use-in-wrestling.html.

140 **These studies and expert recommendations:** Office of Dietary Supplements, National Institutes of Health, "Multivitamin/Mineral Sup-

plements: Fact Sheet for Health Professionals," July 8, 2015, https://ods.od.nih.gov/factsheets/MVMS-HealthProfessional/.

141 **additional studies needed:** Vikas Kapil et al., "Physiological Role for Nitrate-reducing Oral Bacteria in Blood Pressure Control," *Free Radical Biology and Medicine* 55 (February 2013): 93–100, www.ncbi.nlm.nih.gov/pmc/articles/PMC3605573/.

143 **Illustration:** Jessica L. Mark Welch et al., "Biogeography of a Human Oral Microbiome at the Micron Scale," *PNAS* 113, no. 6 (2016): 791–800, http://www.pnas.org/content/113/6/E791.abstract.

144 **The average person:** Allison Aubrey, "The Average American Ate (Literally) a Ton This Year," *The Salt,* NPR, December 31, 2011, www.npr.org/sections/thesalt/2011/12/31/144478009/the-average-american-ate-literally-a-ton-this-year.

157 **Mary Schluckebier, who leads:** Cameron Scott, "Is Non-Celiac Gluten Sensitivity a Real Thing?," Healthline, April 16, 2015, www.healthline.com/health-news/is-non-celiac-gluten-sensitivity-a-real-thing-041615.

169 **in a 2015 review:** Catherine J. Andersen, "Bioactive Egg Components and Inflammation," *Nutrients* 7(9): 7889–7913, www.ncbi.nlm.nih.gov/pmc/articles/PMC4586567/.

171 **He offers glimmers of hope:** I.-J. Wang and J.-I. Wang, "Children with Atopic Dermatitis Show Clinical Improvement After Lactobacillus Exposure," *Clinical and Experimental Allergy* 45, no. 4 (March 19, 2015): 779–87, doi:10.1111/cea.12489.

176 **typified in a 2013 research review:** Jean-Philippe Bonjour et al., "Dairy in Adulthood: From Foods to Nutrient Interactions on Bone and Skeletal Muscle Health," *Journal of the American College of Nutrition* 32, no. 4 (August 2013): 251–63, doi:10.1080/07315724.2013.816604.

177 **Many countries stopped:** Michael F. Holick, "The Vitamin D Deficiency Pandemic: A Forgotten Hormone Important for Health," *Public Health Reviews* 32 (2010): 267–83, www.publichealthreviews.eu/upload/pdf_files/7/15_Vitamin_D.pdf.

178 **the people who recommend:** "Dairy Farms in the US: Market Research Report," IBISWorld.com, February 2016, www.ibisworld.com/industry/default.aspx?indid=49.

179 **On the Creighton University website:** Robert P. Heaney, "What Is Lactose Intolerance?," January 4, 2013, http://blogs.creighton.edu/heaney/2013/01/04/what-is-lactose-intolerance/.

180 **The 2015 panel:** *Scientific Report of the 2015 Dietary Guidelines Committee,* February 2015, http://health.gov/dietaryguidelines/2015-scientific-report/.

183 **Illustration:** "Obesity System Influence Diagram," *Shift* (2008), http://www.shiftn.com/obesity/Full-Map.html.

184 **Two thousand years ago:** "History of Vegetarianism—Plutarch (c. AD 46–c.120)," International Vegetarian Union, www.ivu.org/history/greece_rome/plutarch.html.

185 **"tear into the living flesh":** Howard F. Lyman, *Mad Cowboy* (New York: Touchstone, 2001).

FOUR • DRINKING: The Hydrating Parts

194 **The label adopted:** "Her Debut (1900–1921)," Morton Salt, www.mortonsalt.com/heritage-era/her-first-appearance/.

195 **One of the major producers:** Erika Fry, "There's a National Shortage of Saline Solution. Yeah, We're Talking Salt Water. Huh?," *Fortune,* February 5, 2015. http://fortune.com/2015/02/05/theres-a-national-shortage-of-saline/.

195 **In 2013, a hospital in White Plains:** Nina Bernstein, "How to Charge $546 for a Bag of Saltwater," *New York Times,* August 24, 2013.

199 **This "compulsive" water drinking:** Mary Ann Boyd, "Polydipsia in the Chronically Mentally Ill: A Review," *Archives of Psychiatric Nursing* 4, issue 3 (June 1990): 166–75, www.psychiatricnursing.org/article/0883-9417(90)90005-6/abstract.

199 **Against his will:** Melissa Gill and MacDara McCauley, "Psychogenic Polydipsia: The Result, or Cause of, Deteriorating Psychotic Symptoms? A Case Report of the Consequences of Water Intoxication," *Case Reports in Psychiatry* 2015 (2015), doi:10.1155/2015/846459.

203 **In a few hours:** Richard L. Guerrant, Benedito A Carneiro-Filho, and Rebecca A. Dillingham, "Diarrhea, and Oral Rehydration Therapy: Triumph and Indictment," *Clinical Infectious Disease* 37, no. 3 (2003): 398–405, doi:10.1086/376619.

204 **Patients were made:** Anthony Karabanow, MD, "Cholera in Haiti," Crudem Foundation, http://crudem.org/cholera-in-haiti-2/.

204 **The "magic bullet":** Joshua Ruxin, "Magic Bullet: The History of Oral Dehydration Therapy," *Medical History* 38 (1994): 363–97.

206 **Using the intestine:** Mark O. Bevensee, ed., *Co-Transport Systems* (San Diego: Academic Press, 2012).

207 **Historian David Silbey:** David Silbey, *A War of Frontier and Empire: The Philippine-American War, 1899–1902* (New York: Hill & Wang, 2007).

213 **Coca-Cola clarifies:** Smartwater website, www.drinksmartwater.com.

217 **The entrepreneur J. Darius Bikoff:** Gwendolyn Bounds, "Move Over, Coke," *Wall Street Journal,* January 30, 2006.

217 **The International Bottled Water Association defined:** International Bottled Water Association to Food and Drug Administration, December 23, 2003, www.fda.gov/ohrms/dockets/dailys/03/dec03/122403/02N-0278-C00271-vol21.pdf.

218 **Since bottled water:** "Bottled Water Industry Statistics," Statistic Brain, April 2015, www.statisticbrain.com/bottled-water-statistics.

218 **The transition from quarter waters:** Theresa Howard, "50 Cent, Glaceau Forge Unique Bond," *USA Today,* December 17, 2007.

219 **In 2006, Glacéau founder:** William Neuman, "Liquid Funds for a Penthouse," *New York Times,* April 23, 2006.

219 **A highlight for him:** Tom Philpott, "Coke: Wait, People Thought Vitaminwater Was Good for You?," *Mother Jones,* January 18, 2013, www.motherjones.com/tom-philpott/2013/01/coca-cola-vitamin-water-obesity.

220 **The judge did find:** Susanna Kim, "Court Rules Vitaminwater Lawsuit Can Move Forward," ABC News, July 19, 2013.

229 **Illustration:** Michela Leonardia et al., "The Evolution of Lactase Persistence in Europe. A Synthesis of Archaeological and Genetic Evidence," *International Dairy Journal* 22, no. 2 (2012): 88–97, http://dx.doi.org/10.1016/j.idairyj.2011.10.010.

230 **Hunter S. Thompson:** Juan F. Thompson, *Stories I Tell Myself* (New York: Knopf, 2016).

FIVE • RELATING: The Sex Parts

242 **Only in 1994:** "The Genetics of Sex Determination: Rethinking Concepts and Theories." Gendered Innovations, Stanford University, http://genderedinnovations.stanford.edu/case-studies/genetics.html.

243 **Only in February 2013:** David J. Goodman, "See Topless Woman? Just Move On, Police Are Told," *New York Times,* May 15, 2015.

244 **According to Andreas Hahn:** European College of Neuropsychopharmacology, "Research Shows Testosterone Changes Brain Structures in Female-to-male Transsexuals," August 31, 2015, www.ecnp.eu/~/media/Files/ecnp/About ECNP/Press/AMS2015/Hahn PR FINAL.pdf?la=en.

247 **They write in one academic journal:** Agnieszka M. Zelazniewicz and Boguslaw Pawlowski, "Female Breast Size Attractiveness for Men as a Function of Sociosexual Orientation (Restricted vs. Unrestricted)," *Archives of Sexual Behavior* 40, no. 6 (2011): 1129–35, doi:10.1007/s10508-011-9850-1.

248 **Other researchers have found:** Viren Swami and Martin J. Tovée, "Resource Security Impacts Men's Female Breast Size Preferences," *PLoS ONE* 8, no. 3 (2013): e57623, doi:10.1371/journal.pone.0057623.

249 **For example, breast reduction surgery:** M. Nadeau, et al. "Analysis of Satisfaction and Well-Being Following Breast Reduction Using a Validated Survey Instrument," *Plastic and Reconstructive Surgery* 132, no. 2 (2013): 285–90. "Breast Reduction Surgery Found to Improve

Physical, Mental Well-Being" (news release), July 30, 2013, EurekAlert!, www.eurekalert.org/pub_releases/2013-07/wkh-brs073013.php.

250 **Psychologist Gordon Gallup:** Alan F. Dixson, *Sexual Selection and the Origins of Human Mating Systems* (Oxford: Oxford University Press, 2009).

251 **Marmosets ejaculate:** Ibid.

252 **In a lab, women:** John Heidenry, *What Wild Ecstasy: The Rise and Fall of the Sexual Revolution* (New York: Simon & Schuster, 1997).

254 **In the years that Londa Schiebinger:** Robert Proctor and Londa L. Schiebinger, *Agnotology: The Making and Unmaking of Ignorance* (Stanford, CA: Stanford University Press, 2008).

256 **The glans alone:** Barry S. Verkauf et al., "Clitoral Size in Normal Women," *Obstetrics and Gynecology* 80, no. 1 (July 1992).

256 **Females on the smaller end:** Bahar Gholipour, "Women's Orgasm Woes: Could 'C-Spot' Be the Culprit?," Live Science, February 20, 2014, www.livescience.com/43528-clitoris-size-orgasm.html.

257 **Still, unlike male balls:** Emmanuele Jannini et al., "Beyond the G-Spot: Clitourethrovaginal Complex Anatomy in Female Orgasm," *Nature Reviews Urology,* no. 11 (August 12, 2014): 531–38, doi:doi:10.1038/nrurol.2014.193.

258 **G-spot denier:** Kenny Thapoung, "The Secret to Better Orgasms: The C-Spot?," *Women's Health,* February 25, 2014, www.womenshealthmag.com/sex-and-love/c-spot.

258 **These ideas coalesce:** Jannini et al., "Beyond the G-Spot."

260 **Recent MRI studies:** John Bancroft, *Human Sexuality and Its Problems* (Edinburgh: Churchill Livingstone, 1989).

260 **That Viagra can enhance:** Jennifer R. Berman et al., "Safety and Efficacy of Sildenafil Citrate for the Treatment of Female Sexual Arousal Disorder: A Double-Blind, Placebo Controlled Study," *Journal of Urology* 170, no. 6 (2003): 2333–38, doi:10.1097/01.ju.0000090966.74607.34; "Study Finds Viagra Works for Women," ABC News, April 28, 2016.

262 **"multifunctional serotonin agonist antagonist":** S. M. Stahl, "Mechanism of Action of Flibanserin, a Multifunctional Serotonin Agonist and Antagonist (MSAA), in Hypoactive Sexual Desire Disorder," *CNS Spectrums* 20(1):1–6, www.ncbi.nlm.nih.gov/pubmed/25659981.

262 **The company then convinced:** "Addyi Approval History," Drugs.com, www.drugs.com/history/addyi.html.

262 **People who take Addyi:** Diana Zuckerman and Judy Norsigian, "The Facts About Addyi, Its Side Effects and Women's Sex Drive," Our Bodies Ourselves, September 8, 2015, www.ourbodiesourselves.org/2015/09/addyi-side-effects-and-womens-sex-drive/.

262 **Spontaneous loss of consciousness:** Ibid.

262 **They called the Even the Score petition:** Ibid.

263 **Pfizer publically abandoned:** Gardiner Harris, "Pfizer Gives Up Testing Viagra on Women," *New York Times,* February 28, 2008.

263 **From an epicenter in Australia:** Karmen Wai et al., "Fashion Victim: Rhabdomyolysis and Bilateral Peroneal and Tibial Neuropathies as a Result of Squatting in 'Skinny Jeans,'" *Journal of Neurology, Neurosurgery and Psychiatry* 87, no. 7 (2015): 782, doi:10.1136/jnnp-2015-310628.

264 **The twenty-two-year-old's pelvis:** J. H. Scurr and P. Cutting, "Tight Jeans as a Compression Garment After Major Trauma," *BMJ* 288, no. 6420 (1984): 828, doi:10.1136/bmj.288.6420.828.

268 **Labiaplasties for that purpose:** David Veale et al., "Psychosexual Outcome After Labiaplasty: A Prospective Case-Comparison Study," *International Urogynecology Journal* 25, no. 6 (2014): 831–39, doi:10.1007/s00192-013-2297-2.

268 **According to Nielsen, NYU:** Elisabeth Rosenthal, "Ask Your Doctor If This Ad Is Right for You," *New York Times,* February 27, 2016.

274 **So a person has to:** Lisa Richards, "The Anti-Candida Diet," The Candida Diet, www.thecandidadiet.com/anti-candida-diet/.

275 **The vast majority of syphilis cases:** Centers for Disease Control and Prevention, "Syphilis Statistics," www.cdc.gov/std/syphilis/stats.htm.

276 **Still, more than fifty thousand Americans:** Centers for Disease Control and Prevention, "Table 1. Sexually Transmitted Diseases—Reported Cases and Rates of Reported Cases per 100,000 Population, United States, 1941–2013," www.cdc.gov/std/stats13/tables/1.htm.

SIX • **ENDURING:** The Dying Parts

286 **Even as technology:** Cedars-Sinai Heart Institute, "World Health Organization Study: Atrial Fibrillation Is a Growing Global Health Concern," December 17, 2013, www.cedars-sinai.edu/About-Us/News/News-Releases-2013/World-Health-Organization-Study-Atrial-Fibrillation-is-a-Growing-Global-Health-Concern.aspx.

286 **Atrial fibrillation makes:** Karin S. Coyne et al., "Assessing the Direct Costs of Treating Nonvalvular Atrial Fibrillation in the United States," *Value in Health* 9, no. 5 (2006): 348–56, doi:10.1111/j.1524-4733.2006.00124.x.

286 **The heart malfunction:** Cedars-Sinai Heart Institute, "World Health Organization Study: Atrial Fibrillation Is a Growing Global Health Concern."

287 **More people will die:** World Health Organization, "The Top 10 Causes of Death," May 2014, www.who.int/mediacentre/factsheets/fs310/en/.

289 **Between 2 percent and 16 percent:** Madeleine Stix, "Un-extraordinary Measures: Stats Show CPR Often Falls Flat," CNN, July 10, 2013.

289 **Resuscitation is slightly more effective:** Ibid.

290 **The TV study was important:** Susan J. Diem, John D. Lantos, and James A. Tulsky, "Cardiopulmonary Resuscitation on Television—Miracles and Misinformation," *New England Journal of Medicine* 334, no. 24 (1996): 1578–82, doi:10.1056/nejm199606133342406.

291 **Legal scholars have similarly:** Sirun Rath, "Is the 'CSI Effect' Influencing Courtrooms?," NPR, February 5, 2011.

293 **The year was 1978:** J. Vedel, "[Permanent Intra-hisian Atrioventricular Block Induced During Right Intraventricular Exploration]," *Archives des Maladies du Coeur et des Vaisseaux* 72, no. 1 (January 1979).

294 **Rajappan understatedly calls:** J. P. Joseph and K. Rajappan, "Radiofrequency Ablation of Cardiac Arrhythmias: Past, Present and Future," *QJM* 105, no. 4 (2011): 303–14, doi:10.1093/qjmed/hcr189.

295 **In San Francisco:** R. Gonzalez et al. "Closed-Chest Electrode-Catheter Technique for His Bundle Ablation in Dogs," *American Journal of Physiology—Heart and Circulatory Physiology* 241, no. 2 (August 1981).

296 ***The New York Times* reported in 1998:** Marcelle S. Fisher, "Doctor Serves as an Electrician for the Heart," *New York Times,* June 7, 1998.

296 **Michel Haïssaguerre and colleagues:** Michel Haïssaguerre et al., "Spontaneous Initiation of Atrial Fibrillation by Ectopic Beats Originating in the Pulmonary Veins," *New England Journal of Medicine* 339, no. 10 (1998): 659–66, doi:10.1056/nejm199809033391003.

298 **While some people do:** Andrea Skelly et al., "Catheter Ablation for Treatment of Atrial Fibrillation," Agency for Healthcare Research and Quality, April 20, 2015.

298 **Rather, in 2015:** Henry D. Huang et al., "Incidence and Risk Factors for Symptomatic Heart Failure After Catheter Ablation of Atrial Fibrillation and Atrial Flutter," *Europace* 18, no. 4 (2015): 521–30, doi:10.1093/europace/euv215.

299 **Another in Europe:** "Summary," Early Treatment of Atrial Fibrillation for Stroke Prevention Trial, www.easttrial.org/summary.

300 **The results of his:** H. S. Abed et al., "Effect of Weight Reduction and Cardiometabolic Risk Factor Management on Symptom Burden and Severity in Patients with Atrial Fibrillation: A Randomized Clinical Trial," *Journal of the American Medical Association* 310, no. 19 (2013): 2050–60, doi:10.1001/jama.2013.280521.

301 **Sanders got slightly:** Rajeev K. Pathak et al., "Long-Term Effect of Goal-Directed Weight Management in an Atrial Fibrillation Cohort," *Journal of the American College of Cardiology* 65, no. 20 (2015): 2159–69, doi:10.1016/j.jacc.2015.03.002.

302 **But by 1997:** Rick A. Nishimura et al., "Dual-Chamber Pacing for Hypertrophic Cardiomyopathy: A Randomized, Double-Blind, Crossover Trial," *Journal of the American College of Cardiology* 29, no. 2 (1997): 435–41, doi:10.1016/s0735-1097(96)00473-1.

302 **Indeed, it had:** Michael Doumas and Stella Douma, "Interventional Management of Resistant Hypertension," *Lancet* 373, no. 9671 (2009): 1228–30, doi:10.1016/s0140-6736(09)60624-3.

303 **It causes strokes:** World Health Organization, *A Global Brief on Hypertension* (2013), http://www.who.int/cardiovascular_diseases/publications/global_brief_hypertension/en/.

303 **In the United States alone:** Centers for Disease Control and Prevention, "High Blood Pressure Facts," www.cdc.gov/bloodpressure/facts.htm.

303 **By 2014, doctors:** Deepak L. Bhatt et al., "A Controlled Trial of Renal Denervation for Resistant Hypertension," *New England Journal of Medicine* 370 (April 10, 2014): 1393–401, doi:10.3410/f.718329296.793495177.

303 **That quarter, Medtronic:** Chris Newmaker, "Medtronic Loses $236 Million After Renal Denervation Failure," QMed, February 18, 2014, www.qmed.com/news/medtronic-loses-236-million-after-renal-denervation-failure.

304 **Sanjay Kaul has said:** Larry Husten, "WSJ Attack on Sham Surgery Is About Healthy Profits, Not Patients," *Forbes,* February 20, 2014.

305 **When the World Heart Federation:** World Heart Federation, "World Heart Federation Introduces 'DIY Pulse Test' to Help Fight Against Atrial Fibrillation & Stroke" (news release), October 22, 2012, www.world-heart-federation.org/press/releases/detail/article/world-heart-federation-introduces-diy-pulse-test-to-help-fight-against-atrial-fibrillation-s/.

307 **By 2050, expect:** Fergus Walsh, "Superbugs to Kill 'More Than Cancer' by 2050," BBC News, December 11, 2014.

308 **"If we fail to act":** Fergus Walsh, "Antibiotic Resistance: Cameron Warns of Medical 'Dark Ages,'" BBC News, July 2, 2014.

311 **The team kept analyzing:** "Hereditary Leiomyomatosis and Renal Cell Cancer," Genetics Home Reference, June 21, 2016, https://ghr.nlm.nih.gov/condition/hereditary-leiomyomatosis-and-renal-cell-cancer.

311 **About one hundred families:** Ibid.

317 **In late 2015, researchers:** Food and Drug Administration, "What Are Stem Cells? How Are They Regulated?," May 31, 2016, www.fda.gov/AboutFDA/Transparency/Basics/ucm194655.htm.

320 **In 2013, Sinclair reported:** David Cameron, "A New—and Reversible—Cause of Aging," Harvard Medical School, December 19, 2013, https://hms.harvard.edu/news/genetics/new-reversible-cause-aging-12-19-13.

321 **"If this paper is right":** Ed Yong, "Clearing the Body's Retired Cells Slows Aging and Extends Life," *The Atlantic,* February 13, 2016.

327 **Illustration:** Chris Morley, "How to Improve Your Posture Infographic," *Chris Morley Design,* accessed September 22, 2016, http://chrismorleydesign.com/portfolio-items/how-to-improve-your-posture-forever-infographic.

333 **The median cost of embalming:** National Funeral Directors Association, "Trends in Funeral Service," www.nfda.org/newstrends-in-funeral-service.

333 **"Losing a loved one":** "I Need a Loan," First Franklin Financial Corporation, www.1ffc.com/loans/#.V4gGoY54O8Y.

333 **The burial "vault":** National Funeral Directors Association, "Trends in Funeral Service."

334 **Even Walmart:** "The Official Vatican Observatory Foundation Mahogany Casket, Sacred Heart II," Walmart, www.walmart.com/ip/The-Official-Vatican-Observatory-Foundation-Mahogany-Casket-Sacred-Heart-II/38042564.

334 **Cotton placed in the anus:** Funeral Consumers Alliance, "Embalming: What You Should Know," www.funerals.org/what-you-should-know-about-embalming/.

337 **But as a biology lab:** K. Kelvin, P. Lim, and N. Sivasothi, "A Guide to Methods of Preserving Animal Specimens in Liquid Preservatives," 1994, http://preserve.sivasothi.com.

337 **Beyond its toxicity:** "Formaldehyde: Toxicology," Carcinogenic Risk in Occupational Settings (CRIOS), www.crios.be.

337 **To make matters worse:** Joann Loviglio, "Kids Use Embalming Fluid as Drug," ABC News, July 27, 2014, http://abcnews.go.com/US/story?id=92771.

338 **Or for $4.95:** "Coffin Plans to Make Your Own Plywood Coffin," Piedmont Pine Coffins, http://piedmontpinecoffins.com/diy-coffin-plans/.

339 **In a process:** Eric Spitznagel, "The Greening of Death," *Bloomberg Businessweek,* November 3, 2011, www.bloomberg.com/news/articles/2011-11-03/the-greening-of-death.

340–1 **Illustration:** Chuck Lakin, "Coffin Plans," *Last Things* (blog), http://www.lastthings.net/coffin-plans.

343 **If there are ethical:** David Barboza, "China Turns Out Mummified Bodies for Displays," *New York Times,* August 8, 2006.

346 **The highly effective:** "About the USPHS Syphilis Study," Tuskegee University, www.tuskegee.edu/about_us/centers_of_excellence/bioethics_center/about_the_usphs_syphilis_study.aspx.

SELECTED BIBLIOGRAPHY

Biss, Eula. *On Immunity: An Inoculation*. Minneapolis: Graywolf Press, 2014.

DeSalle, Rob, Susan L. Perkins, and Patricia Wynne. *Welcome to the Microbiome: Getting to Know the Trillions of Bacteria and Other Microbes In, On, and Around You.* New Haven, CT, and London: Yale University Press, 2015.

Dobscha, Susan. *Death in a Consumer Culture*. New York: Routledge, 2016.

Epstein, David J. *The Sports Gene: Inside the Science of Extraordinary Athletic Performance*. New York: Current, 2013.

Knight, Rob, and Brendan Buhler. *Follow Your Gut: The Enormous Impact of Tiny Microbes*. New York: Simon & Schuster/TED, 2015.

Levinovitz, Alan. *The Gluten Lie: And Other Myths About What You Eat*. New York: Regan Arts, 2015.

Mukherjee, Siddhartha. *The Gene: An Intimate History.* New York: Scribner, 2016.

Pollan, Michael. *The Omnivore's Dilemma: A Natural History of Four Meals*. New York: Penguin Press, 2006.

Price, Catherine. *Vitamania: Our Obsessive Quest for Nutritional Perfection*. New York: Penguin, 2016.

Skloot, Rebecca. *The Immortal Life of Henrietta Lacks*. New York: Crown, 2010.

Washington, Harriet. *Infectious Madness*. New York: Little, Brown, 2015.

ABOUT THE AUTHOR

JAMES HAMBLIN is a writer and senior editor at *The Atlantic* magazine. He hosts the video series "If Our Bodies Could Talk," for which he was a finalist in the 2015 Webby Awards for Best Web Personality. He is the recipient of a 2015 Yale University Poynter Fellowship in journalism. His work has been featured in/on *The New York Times, Politico,* NPR, BBC, MSNBC, *New York, The Awl,* and *The Colbert Report*. *Time* named him among the 140 people to follow on Twitter in 2014. He's based in Brooklyn, New York.